William James, MD

William James, MD

PHILOSOPHER, PSYCHOLOGIST, PHYSICIAN

Emma K. Sutton

The University of Chicago Press   *Chicago and London*

The University of Chicago Press, Chicago 60637
The University of Chicago Press, Ltd., London
© 2023 by Emma K. Sutton
Published 2023
Printed in the United States of America

32 31 30 29 28 27 26 25 24 23   1 2 3 4 5

ISBN-13: 978-0-226-82896-1 (cloth)
ISBN-13: 978-0-226-82898-5 (paper)
ISBN-13: 978-0-226-82897-8 (e-book)
DOI: https://doi.org/10.7208/chicago/9780226828978.001.0001

This book is produced with the generous assistance of a grant from the
Isobel Thornley Bequest to the University of London.

Library of Congress Cataloging-in-Publication Data

Names: Sutton, Emma K., author.
Title: William James, MD : philosopher, psychologist, physician / Emma K.
  Sutton.
Description: Chicago : The University of Chicago Press, 2023. | Includes
  bibliographical references and index.
Identifiers: LCCN 2023008540 | ISBN 9780226828961 (cloth) |
  ISBN 9780226828985 (paperback) | ISBN 9780226828978 (ebook)
Subjects: LCSH: James, William, 1842–1910. | Physicians—United States—
  Biography.
Classification: LCC B945.J24 S88 2023 | DDC 109.2 [B]—dc23/eng/20230427
LC record available at https://lccn.loc.gov/2023008540

♾ This paper meets the requirements of ANSI/NISO Z39.48-1992
(Permanence of Paper).

For Richard, Albert, and Josiah

CONTENTS

FIGURES

The Public Physician

Diagnosing James

Describing William James (1842–1910) has never been a straightforward task. While he was alive he wandered at will across several academic disciplines and acquired, over the course of his Harvard career, posts in physiology, philosophy, and psychology.[1] During the late nineteenth and early twentieth centuries he published texts, such as *The Principles of Psychology* (1890), *The Will to Believe* (1897), *The Varieties of Religious Experience* (1902), and *Pragmatism* (1907), which range in subject matter from experimental psychology and metaphysics to the psychology of religion and the philosophy of knowledge.[2] His metaphorical trophy case glitters with an impressive and eclectic array of accolades and titles. He was described as "the psychological pope of the new world" and after his death as "America's foremost philosophical writer"; a man whose passing "leaves vacant a place in the world of English letters which no living writer and thinker can fill"; and even a "sage and saint" from whom religious truth "constantly shone out . . . shone, as it were, straight through his waistcoat and distributed itself to everyone in the drawing-room, or in the lecture-hall where he sat."[3]

There is one element of James's life and work that unites these disparate identities, however. In 1869, several years before he secured his first lectureship, he graduated from Harvard Medical School and earned his MD. Hampered by his own ill health, James abandoned his plans to practice as a doctor, but these studies were only the beginning of a profound and lifelong occupation with questions about the essential nature of health, healing, and invalidism and their implications for society. His writings, across their disciplinary breadth, return time after time to issues of a medical provenance. In this book I make the case that James's

0.1 James was awarded his MD from Harvard Medical School in March 1869, after more than five years of interrupted study. This certificate lists his examiners, who included Oliver Wendel Holmes Sr., and the subject of his thesis, namely, the effects of cold on the body. (Diplomas, degrees, notifications of appointments, etc., William James papers [MS Am 1092.9–1092.12, MS Am 1092.9 (4571), Box: 40], Houghton Library, Harvard University.)

medical interests, concerns, and values are the threads that bind many of his seemingly unconnected pursuits together. They are the warp and weft of many of his best-known publications and major lines of thought.[4]

This medical focus was a product not just of James's own training, but also of the broader social milieu in which he moved. He enjoyed friendships with junior members of the Bowditch and Holmes families whose ranks included Henry Ingersoll Bowditch (1808–92) and Oliver Wendell Holmes Sr. (1809–94), two of the most eminent American physicians of the nineteenth century who were responsible for significant reforms in public health planning and general medical practice, respectively. Bowditch wrote a seminal text on *Public Hygiene in America*, and Holmes coined the term "anaesthesia," a concept that would later find its way into the heart of one of James's earliest essays on materialism and religious faith.[5] James's close colleagues at Harvard numbered Henry Pickering Bowditch (1840–1911), who studied alongside him and later went on to become dean of the medical school, and James Jackson Putnam (1846–1918), who was a founding member of the American Neurological Association. Further afield, James corresponded with many of the leading medical psychologists, alienists, and neurologists from across America, continental Europe, and Britain.

In addition to these representatives of the professional elite, the English doctor James John Garth Wilkinson (1812–99) was a respected associate of James's father, Henry James Sr. (1811–82), and a vociferous proponent of homeopathy and other unorthodox medical ideas.[6] During the 1880s James himself began a long-term collaboration with members of the Society for Psychical Research, whose observations and theories he found stimulating with respect to his own ideas about health and healing.[7] Living in Cambridge, Massachusetts, James was, moreover, at the geographical epicenter of the American mind-cure or lay mental healing movement. When this movement gathered considerable popular momentum during the 1880s and 1890s, he encountered the feats of these alternative practitioners not as abstract accounts or dismissible hearsay, but rather as they rippled through his social circle animating the lives of friends and family in a fashion that he could not ignore.

In this way, James's life was peopled with the key protagonists in some of the most significant medical controversies and developments of the day. Furthermore, these debates were not merely of relevance to those plying a physician's trade of one denomination or another. James lived in a late nineteenth-century world that witnessed the rise of a newfound authority for medical science, an authority that reached outside the confines of the doctor-patient encounter and into the daily lives and thoughts of his contemporaries.[8] The tendrils of novel medical prac-

tices and perspectives curled themselves around the religious and political discussions of the period with far-reaching consequences. The mid-
century deployment of the earliest effective anesthetics such as ether, for
example, was seen as a potent sign of medical power; these pain-relieving
revelations disrupted long-standing religious narratives that positioned
suffering as a punishment from God or a path to redemption.[9]

Equally, in the latter decades of the century, the establishment of state
boards of health and their legislative influence instantiated a new emphasis on personal health as a public concern. The resulting investigations and sanitary reforms were widely welcomed, but the philosophy
behind some public hygiene rhetoric was felt by James to be less benign.
Talk of the "political economy of health" brought the statistical meth-
odologies and materialist value system of economics to bear on the assessment of the merits of an individual human life. Those who endured
infirmity and disease were depicted wholly as a burden: a financial drain
on the resources of the state that must carry the cost of supporting these
less productive citizens.[10]

Meanwhile, evolutionary ideas about health, heredity, and the ideal
of adaptation to the environment, in the work of Herbert Spencer
(1820–1903), Charles Darwin (1809–82), and others, had become the
basis for a variety of body-centered ethics and fears. Old assumptions
about faith and the moral and religious possibilities for human worth
and redemption were increasingly displaced by a pervasive concern with
the medically "normal" and the perfectibility of the body, rather than
the soul.[11] In its most extreme form, this biological mandate manifested
itself in treatises warning of the degeneration of the human race and
comprised the origins of the eugenics movement.[12] In short, there was
a growing social stigma attached to illness and invalidism. This was a
world in which James feared that medical diagnoses had become "clubs
to knock men down with."[13]

James felt himself, moreover, to be personally implicated in these disturbing intellectual currents. Most chronicles of his life have portrayed
a distressed young man, a psychological or spiritual crisis, followed by
the emergence of a mature thinker who threw off his pallor of mental
sickness for good.[14] In contrast, I argue that James considered himself
a genuine invalid to the end of his days. Consequently, he was far from
being a disinterested observer of the medico-ethical ideas under discussion. Rather these were issues that took on a pronounced private
significance for him and were, I contend, partly for this reason deeply
integrated into his thinking throughout his career.

Although the majority of James's accounts of his ill health have received scant attention, his youthful troubles dating to the brief "crisis"

period during the late 1860s and early 1870s have been considered at length. From "soul sickness" to repressed rage, unconscious sexual longings, "depression," and "bipolar mood disorder," the underlying causes of James's symptoms have been analyzed from within a variety of contemporary psychological and philosophical schemas.[15] These retrospective diagnoses can perform a useful biographical service, bringing James and his experiences closer to our own worlds of understanding. In this book, however, I take the opposite approach: I attempt to move us into more familiar contact with James's nineteenth-century universe of medical meanings. I argue that the consequences of James's *self*-diagnoses, and the significance he attached to them, extended beyond the immediacy of his own suffering and into the metaphysical, ethical, psychological, and political content of his work.

A Philosophy of Everyday Life

James was not one to suffer in silence. He complained frequently and in detail about the state of his health to family and friends. His grumbles about his eyes, his back, his insomnia, his low spirits, and even his bowels are so commonplace that it's tempting to approach them with an attitude of lighthearted dismissal. I take James's reports of his symptoms and prognoses more seriously, however. He explicitly and habitually framed these medical confidences with reference to a range of philosophical topics and arguments, themes that subsequently manifested themselves in his public writings.

James's melancholy opened up questions about the relationship between the mind and body; his pain was presented and probed as a form of metaphysical evil; the crippling nature of his back condition was positioned as an ethical threat to his ability to contribute to society; this combined burden of invalidism represented a moral embargo on fatherhood with its risk of passing on a sickly inheritance; and, throughout his life, he prized religious faith, first and foremost, as a stimulus or tonic for those struggling with illness and infirmity. Wherever you look, James's corpus is riddled with disease.

The consideration I give to the medical minutiae of James's life and letters is, moreover, part of a more general methodological commitment. Underlying this book is the question of how and why people come to acquire and champion the particular ideas they profess. The conventional historical answer to this question is that context is key, but contexts come in many different shapes and sizes. Traditionally, historical scholarship has tended to focus, partly out of practical necessity, on a relatively limited set of macro-contexts when constructing the intellec-

tual trajectory of individuals and groups, drawing attention to the implications of momentous events, socioeconomics, and the cultural significance of landmark publications and influential figures.[16] The breadth
of James's literary remains, however, make it possible to conceive of
a multilayered exploration of his philosophical development, one that
pays considerable attention to the micro-contexts of everyday life.

I draw upon my reading of the approximately 9,400 extant letters
written either by or to James, and his unpublished notebooks, diaries,
and reading lists, to show how, for him, philosophizing and oratory
were not abstract or lofty occupations, but impassioned responses to
his own ordinary life experiences and challenges.[17] His philosophy was
not something he picked up and put down when he entered or left his
study; he was just as likely to carry it with him into his water closet. As
unseemly as it may sound to some ears, James was as capable of starting a metaphysical discussion about his own and his brother's constipation as he was about Kant.[18] I argue that James's thinking was formed in
reaction not just to the academic provocations of his colleagues, or to
large-scale historical developments, the wars and general machinations
of government, but also to the intimate details of his personal world and
especially the mundane machinations of his own body.[19]

This is not to suggest, as did James's first, official biographer, that the
state of his health determined his thinking in any simple causal way.
Philosopher Ralph Barton Perry (1876–1957), in his influential two-
volume account of James's life and thought mentioned, in passing, "certain well-marked prepossessions" that guided James as he began to nav-
igate his own philosophical craft, stating that these prepossessions were
determined, at least in part, by "the condition of his health" and "cannot be traced to any philosophical teaching."[20] In this way, Perry introduced a philosophical distance between his subject's body and mind
that cuts against the whole thrust of James's own interest in the mind-
body enigma and his pragmatic philosophy. In his lectures on *Pragmatism*, James stressed the importance of the human element in knowledge building. He argued that our empirical observations are not inert
with regard to our own investigations; the world does not sit patiently
still while we poke and prod it. Rather our experiences of the world are
actively shaped by the particular interests and type of attention that we
bring to them. Reality, including the reality of our own bodies, is, to a
significant extent, a product of our own making.[21] Similarly, I do not
mean to imply that James thought the way he did because he became
physically unwell, but instead that his sense of himself as unwell and his
philosophical theories were part of an ongoing and intellectually fertile
Jamesian dialogue. His construal and experience of illness and health

were in flux throughout his life. These were fluid concepts, not clearly defined physical facts that led him to favor certain ideas over others.

Indeed, James spent most of his career resisting the notion that our bodies determine the course of our lives as if we are nothing but fleshy machines, a school of thought that he referred to as "medical materialism."[22] He constantly worried at the assumptions of this doctrine, which many of his contemporaries found persuasive. In particular, the question of where the cause of our suffering truly resides, whether in our bodies, our minds, or an altogether different, metaphysical realm, was one that greatly occupied James. Over the course of his lifetime, he interpreted his own and others' symptoms within an eclectic assortment of models. These included inflamed tissues, nervous exhaustion, bad habits, buried emotions and pathological "fixed ideas," inadequate self-discipline, and a "divided self," and he explored a similarly diverse array of preventive and curative practices. Crucially, these Jamesian remappings of the medical terrain were often accompanied by a realigning of his epistemological, ethical, and metaphysical loyalties. I argue that James's attitude toward different facts, inquirers, and beliefs changed considerably over the course of his lifetime, and these philosophical transitions and transformations were rooted within his distinctly medical purview.

This dynamic depiction of James is a defining feature of my analysis. I eschew the assumption that there is any kind of "essential" or "mature" James that may be distilled from his writings. Nor do I assume that the years of his youth were somehow more formative than those of his middle or later decades.[23] Instead, I treat him as a moving target, someone who continued to revise their intellectual positions throughout their life. A vital feature of this book is the weight I give to relevant events, associations, and sources from his private life throughout the length of his career: his marital doubts, the death of his son, his brother's alcoholism, his numerous appointments with spiritual healers, his intense preoccupation with his levels of nervous energy, his midlife melancholy, and his struggle to come to terms with a heart condition and his own mortality. I bring to light a rich and *lifelong* interweaving of personal experience and intellectual development.

This alertness to the shifting nature of James's values and ideas is another consequence of my interest in the everyday nature of his philosophy. I have made a deliberate choice to bring attention to the micro-*texts* as well as the micro-*contexts* of his developing thought; I do not privilege his canonical works as superior instances of Jamesian reflection but instead treat all his writings as evidence of the philosopher in action. His book reviews, manuscript notes, diaries, and extensive correspondence capture the twists, turns, and even reversals in James's thought, in the moment, on a day-to-day basis, unlike his published texts, which emerged on a much slower, year-by-year timetable.[24]

This additional level of detail is not the only feature of these micro-texts that earns them their place in the historical record. A letter to a friend may be less polished and thought-through than a published set of lectures but is also, perhaps, a more frank and open statement of the ideas in question and, critically, the private passions that drove them. My account of James is an avowedly emotional one. He himself made the case that philosophical systems owe their existence, in part, to "the desire for a solid outward warrant for our emotional ends."[25] This observation was part of a stronger claim, moreover. James was convinced of the "ubiquitousness of emotional interests in the mind's operations," and several of his writings explicitly challenge the traditional assumption that thinking and feeling may be treated separately.[26] His own emotional motives and reactions, which he regularly acknowledged, would seem then to be a valid and important area of interest for anyone seeking to understand James's ideas on his terms.

Finally, it should be noted that entering James's thought-world requires an awareness of his historical and, at times, idiosyncratic use of certain words and phrases. Speaking Jamesian is the requisite first step toward deciphering the written traces of his thinking that he left behind. At first glance such a task seems beguilingly simple; but beneath James's lucid and engaging prose, a linguistic irreverence lies in wait to dupe the uninitiated. James played with language, stretching his use of terms such as "God," "religion," and "evil" beyond the bounds of their more conventional definitions either then or now. Nor did his own descriptions of these concepts remain constant throughout his life. In one sense then, this book performs the role of a Jamesian dictionary, highlighting and translating the ever-changing language of its subject.

As mercurial as James was in many ways, however, there was also a consistency to his theories and beliefs and the words that he used to express them, namely, the medical agenda within which he put them to work. As he journeyed across the disciplinary landscapes of physiology, psychology, and philosophy, James mined them all for useful insights into a linked set of concerns: the promotion of health; the prevention and amelioration of disease and suffering; and the justification of the place of the invalid within society.

His inquiries into these matters were far from predictable. He entered the hallowed halls of religion in order to fathom the secrets of its "biological function"; visited the theater of war for what it reveals to us about the limits of human endurance; and looked to the vivisection laboratory for inspiration about the ethical significance of human pain and illness. Ultimately, however, what mattered to James above all else was the practical efficacy of the ideas he unearthed on his travels and, to a significant extent, their value to the general, nonacademic audience to whom many of his writings were addressed. James's intellectual pursuits were born of everyday experiences and frequently offered as solutions for the everyman.[27] He may never have practiced as a doctor in the traditional sense, but throughout his life James remained, at heart, a public physician.

Although loosely chronological, this book does not follow a conventional birth-to-death biographical narrative and instead is structured thematically, with its five chapters focusing on a set of different but interconnected topics. In chapter 1, "Misery and Metaphysics," I explore James's early adulthood, his ill health, and the considerable intellectual emphasis he placed on the concept of pain during this period. I examine how James's first attempts at finding his philosophical feet were caught up in his efforts to accept himself as "crippled," "an invalid," a physical "wreck."[28] Specifically, I consider James's examination and framing of themes such as evil, utilitarianism, suicide, and his changing position with regard to physicalist theories of mind and mental disorder. I read James's subsequent antideterminism and antimaterialism as a reaction against the contemporary doctrine of "medical materialism," a medical philosophy that, taken to its ultimate conclusions, expelled any opportunity for the invalid James to either effect his own improvement or posit a worthwhile life for himself.

Chapter 2, "Health and Hygiene," comprises a study of James's youthful interest in physiology and his personal commitment to the hygienists' "laws of health." I demonstrate that, despite his reservations about the project of medical materialism, James remained wedded, along with his peers, to the worship of health as an ethical ideal. This ideal underpinned his public contributions to the contemporary debate on "the alcohol question" and his private deliberations about the advisability of marriage for himself and his brother. This reverence for the goal of personal and public health persisted throughout James's career and structured publications such as his *Talks to Teachers on Psychology* (1899) and "The Gospel of Relaxation" (1899). I show how these writings and others, such as James's *Principles of Psychology*, evidenced his faith in preventive forms of practice that aimed at promoting a healthy mind and nervous system. Some of James's best-known writings on the topics of habit and the emotions were motivated by this fascination with the potential of mental hygiene, long before the establishment of the mental hygiene movement in the twentieth century.

Chapter 3, "Religion and Regeneration," traces James's move away from a scientifically orthodox approach to medical knowledge toward an epistemology that also encompassed mystical experiences and frameworks. I propose that over the course of the second half of his life his philosophical beliefs changed considerably, and I explore how James's willingness to concede increasing authority to religious viewpoints related to his introduction to the *therapeutic* potential of mystical inquiry and the concept of the subconscious self. This narrative links James's changing ideas, and ultimately his lectures on *Pragmatism*, to his interactions with the Society for Psychical Research, his personal encounters with the spiritualist medium Leonora Piper, and his involvement with the American mind-cure movement. During this time, James also became familiar with the work of the French medical psychologist Pierre Janet (1859–1947). Janet's innovative clinical research deployed the notion that one person may be split into several, secondary personalities, and he located the cause of his patient's symptoms in pathological memories, buried in these subconscious, or hidden, selves. James was captivated by these pre-Freudian concepts of the psychological etiology of disease, and I argue that, for him, they provided critical theoretical validation of the empirical results of the faith-healing community.

Chapter 4, "Energy and Endurance," investigates how James's fascination with the concept of human energies, and their implications for health and healing, animated his ethical, religious, and psychological writings throughout his life. I delve into the biographical circumstances in which *The Varieties of Religious Experience* lectures were composed and show how James's perception of his heart problems and of himself as mortally ill shaped their structure and focus. According to James's psychopathological model of "the divided self," which he set out in *The Varieties*, a religious conversion experience is best understood as a healing, energizing process. I trace the thematic continuities between these lectures and other writings, such as his essays on "The Energies of Men" (1907) and "The Moral Equivalent of War" (1910), where he explores other, nonreligious sources of medicinal energy.

Finally, in chapter 5, "Politics and Pathology," I analyze James's political interests and activities and identify them as a "politics of pathology," concerned primarily with the social categories of "the invalid" and "the degenerate." I track his two-pronged attempts, in his "Exceptional Mental States" lectures (1896), *The Varieties of Religious Experience*, and elsewhere, to renegotiate perceptions of society's sick members. These included, first, his efforts to destabilize pejorative claims about the boundaries of mental health and illness, and, second, his efforts to summon a life of meaning and even heroism for the invalid, via his invocation of a moral or religious dimension to the universe.

I also scrutinize James's contributions to more obviously political medical causes, including his outspoken opposition to the proposed state regulation of lay mental healers and, in the last years of his life, his public support for Clifford Whittingham Beers's (1876–1943) national campaign to address the treatment of the insane in mental asylums.

In "Conclusion: Afterlife," I draw together the multiple strands of James's medical morality and the intellectual legacy that he left behind. Those who knew him and his writings testified to his staunch pursuit of healing wisdom for himself and society at large. This goal was the foundation of his personal ethical program, and throughout his life he remained convinced that *"health* is the only good."[29] James's conceptualizations of health, illness, and invalidism were not static, however, and nor were they conventional. He actively subverted nineteenth-century mentalities that anchored the promise and worth of human life within an exclusively biological or socioeconomic materialism. I suggest that, for James, the essence of health was to be found not in objective assessment but subjective experience. It was, above all else, a feeling state, an attitude, an energy: a stubborn metaphysical refusal to submit to the misery of evil.

CHAPTER 1

Misery and Metaphysics

A Dark Business

There was one experience, above all others, that defined the years of James's early adulthood. It contorted and tormented his personal writings, his plans and life path and, most notably, his own body. Throughout the late 1860s and early 1870s, James's correspondence, diary, and notes return, again and again, to the subject of pain. This single topic preoccupied James as he strived to cope with the implications of a debilitating back condition and its associated symptoms. Writhing in pain's grip, he repeatedly debated, with himself and others, how he should manage it, how it was managing him, and, ultimately, what it all meant.

Although, for James, his anguish was rooted firmly in the domain of the corporeal, he became profoundly disturbed by the way its presence extended deep into the realms of ethics and metaphysics. In late nineteenth-century America, personal suffering was a topic of considerable cultural concern. One fellow philosopher and friend of James's, Charles Sanders Peirce (1839–1914), went so far as to describe it retrospectively as "the Age of Pain."[1] Questions about the significance and role of suffering in society animated political, philosophical, and medical discussions of topics as diverse as slavery, evolutionary science, and surgery.[2]

In this last field the discovery and rapid adoption of ether and other anesthetics, during the middle years of the century, cast pain in a new light. The earthly power of anesthesia reconfigured pain as a "treatable pathology" and disturbed established religious connections between suffering and its moral meanings of sacrifice, punishment, trial, and redemption.[3] It was in this context of widespread ethical and metaphysical unrest that James's own chronicle of misery began.

According to James, his back condition, attributed by his mother to

1.1 James was a keen artist and took lessons with the American painter William Hunt for a brief period in his youth. These drawings of cadavers and an unknown figure are taken from one of James's medical school notebooks. (William James drawings [MS Am 1092.2 (44), Box: 1], Houghton Library, Harvard University.)

long hours of toil at the medical school dissection tables, was the reason for him temporarily abandoning his studies in early 1867. To his close friend, Thomas Wren Ward, he wrote later that year that although "the damned thing showed at first a very strong tendency to disappear after repose," his back pain had soon become an intractable fixture. Hence, he explained, his abrupt departure for Germany in April.[4] The health-

giving lands of Europe were, in the eyes of nineteenth-century Bosto-
nians, a medical utopia, with the mineral springs of various spa towns
a particular attraction.

The long months that followed were a ceaseless merry-go-round of
cure-seeking for James: rest in Dresden; medicinal bathing, douching,
mineral water drinking, mud baths, and "an *exclusive* diet of curds &
whey" at the "Fürstenbad" spa in Teplitz; too much walking in Berlin
that "rather undid the benefit" he had got from Teplitz; hence more rest
and another trip to the baths there that did him "no good but rather the
reverse," as he apprised his sister, but was followed, nevertheless, by a
return visit (after a bit more rest in Dresden), to see whether a "3rd *mild*
experiment with Teplitz" might be successful.[5]

Not content to rely solely on the rejuvenating powers of rest and mineral waters, James left no therapeutic stone unturned in his quest for
improvement. A letter to his brother, the writer Henry James, who suffered similar back problems, attests to James's willingness to endure the
punishing trials of more conventional "heroic" remedies as they were
known. The process of raising clusters of blisters on the skin was a common recourse for conditions thought to involve inflammation:

> I have been trying blisters on my back and they do undeniable good.
> Get a number about the size of a 25 cents piece, or of a copper cent.
> Apply one every night on alternative sides of the spine over the dis-
> eased muscles. In the morning prick the bubbles, and cover them
> with a slip of rag with cerate, fastened down by cross straps of stick-
> ing plaster. Try a dozen in this way at first. Then wait two weeks and
> try ½ dozen more,—two weeks, ½ doz more & so on. If the blister-
> ing is done too *continuously* it loses its effect.[6]

Blisters were provoked with the aim of drawing out fluids from the
inflamed part of the body, in accordance with the contemporary theory
of "counter irritation," or the "production of an artificial or secondary
disease, in order to relieve another primary one."[7] The application of
iodine was used to a similar effect, and James also tried this course of
treatment on himself and later recommended it to his brother Robertson (Bob) James when he fell prey to the "family peculiarity":

> I think you will do well for a few times to paint the painful spots with
> iodine. Be very careful not to put on too many coats at once, for the
> pain does not start up immediately and when it does it may be intol-
> erable if you have painted too thick. The best way is to try a couple of
> coats, then wait a ¼ hour; then if it pains but slightly put on another

&c, until the pain becomes right smart. Never paint a larger surface than twice the palm of your hand at once. After a few days the skin begins to peel off, when you must stop your operations on that spot.[8]

Despite James's best efforts, however, his own symptoms continued to bedevil him, with dispiriting relapses following every episode of seeming respite. A growing sense of frustration is a palpable presence in his correspondence. After nine months of dedicated convalescence in Europe he wrote to Bob: "This accursed thing in my back has now lasted for 13 months. It scatters all my plans for the practice of medecine [sic] to the winds, wh. has been a great disappointment to me, inasmuch as I was getting very much interested therein."[9] And, although heartened by Henry's apparent recuperation, subsequent letters suggest a waning of James's confidence in his professional mastery of their shared condition: "I am very glad to hear of your 'lifting cure,'" he wrote to Henry, adding: "It is strange that when the contracting of those muscles in walking &c. should be deleterious, it should be advantageous in this other motion. What a dark business it all is."[10]

This "business" was set to become darker still. By January of 1869 James had come to the reluctant conclusion that he "must not only drop exercise, but also mental labor," since, as he explained to his friend and fellow medic Henry Pickering Bowditch, "it immediately tells on my back." The consequences were disquieting: "I have consequently made up my mind to lose at least a year now in vegetating and doing nothing but survive." Toward the end of the year, with still no permanent cure in sight, James experimented with the therapeutic potential of electricity, explaining to Bowditch that he had been reading the work of an Austrian neurologist in this field, Moriz Benedikt, and had tried "mounting 30 bunsen cells" to his back.[11] To his brother, Henry, he wrote: "I have been galvanizing my back of late but so far ineffectually; or rather I suspect with a bad effect, the left side, which for the sake of comparison I have alone treated, being now more sensitive than the right." He concluded, three weeks later, "Strong galvanism has done me no good."[12]

Meanwhile, even as James exhausted his medical arsenal, his suffering continued to worsen. As 1869 drew to a close, he reported to his brother that "the pain has spread within 3 months into my upper back & neck."[13] By then back home in Cambridge, James confessed to Bowditch: "I have been a prey to such disgust for life during the past 3 months as to make letter writing almost an impossibility. . . . My own condition I am sorry to say goes on pretty steadily deteriorating in all respects."[14] At this stage in his life, James described himself as "crippled" and "'a mere wreck,' bodily."[15]

The "tedious egotism" of the invalid life weighed heavily on James. When "your 'conscience,' in the form of an aspiration towards recovery, rebukes every tendency towards motion excitement or life, as a culpable excess," he explained to his friend Oliver Wendell Holmes, "the deadness of spirit thereby produced, 'must be felt to be appreciated.'"[16] In the spring of 1870 he wrote to his brother Henry, confiding that he felt "melancholy as a whip-poor-will" and that "sighs are hard to express in words."[17] Even the opportunities for solace and distraction that he found in vigorous walks, company, and reading were dictated by the capricious and exacting demands of his back problem. Empathizing with his brother Bob's own situation, a month earlier he wrote:

> It's a bad circle to move in where you are now, I know—sedentary life makes you weak & nervous and exercise hurts your back. I imagine that a little exercise—*never* enough to bring on the pain is beneficial—but it is mighty hard always to keep ahead of the pain— one's natural tendency being to walk *until* the pain is felt, and then stop—which is too late.[18]

It is no exaggeration to say that, for a number of years, James felt his life to be dominated by the experience and expectation of pain. Whether the self-inflicted suffering of his harsh therapeutic regime or the unpredictable agony of his symptoms, it disciplined his waking moments. His long months of "sickness and solitude" traveling alone around Europe left him plenty of time for contemplation, and his thoughts soon turned to the meaning and possible purpose of his afflictions. During this period James developed a metaphysics of misery that was to influence his philosophical thinking acutely throughout his career.

The Problem of Evil

Many of James's writings raise "the problem of evil" as being at the heart of any system of metaphysics. Whereas some philosophies of life are content to ignore them, "evil facts," he insisted, "are a genuine portion of reality."[19] Discussions of the obstinate presence of evil, and its implications, are a feature of his essays "Rationality, Activity and Faith" (1882) and "The Dilemma of Determinism" (delivered in 1884); his address to students on the question "Is Life Worth Living?" (delivered in 1895); his Gifford Lectures on *The Varieties of Religious Experience* (delivered in 1901–2); his Hibbert Lectures on *A Pluralistic Universe* (delivered in 1908); his lectures on *Pragmatism* (delivered in 1906); and his final, unfinished work, *Some Problems of Philosophy* (published posthumously in

1.2 A portrait of James taken in Boston, circa 1869. (Letters to William James from various correspondents and photograph album [MS Am 1092 (1185), 12, Box: 12], Houghton Library, Harvard University.)

1911).[20] "The problem of evil" was not, for James, merely an abstract philosophical phenomenon. He was particularly troubled by its existence and significance because he came to believe that it lived inside of him.

James had not always felt this way about evil, however. His young, sixteen-year-old self was convinced that it was a meaningless construct, or institutional by-product, avoidable in the event that society ceased to police itself according to its current system of rules. In 1858 he wrote to a friend: "Yes, mankind is all good and as soon as governments and priests are abolished, such a thing as sin will not be known. There is no such thing as a bad heart. . . . All the evil in the world comes from the law and the priests and the sooner these two things are abolished the better."[21] It would appear to be this kind of early thinking that he subsequently described as antinomian or disregardful of moral law. Several years later, in a letter to his friend Ward, he explained how:

> I have grown up partly fm. education & the example of my Dad, partly I think from a natural tendency in a very non-optimistic view of Nature, going so far as to have some years ago a perfectly passionate aversion to all moral praise &c &c—an anti-nomian tendency in short. I have regarded the affairs of human life to be only a phantasmagory, wh. had to be *interpreted* elsewhere in the Kosmos into its real significance.[22]

A survey of Henry James Sr.'s writings on evil makes it possible to elaborate on James's comments. In his work on *The Nature of Evil* his father explained that all actions, whether of apparent moral goodness or evil, are derived directly from God, hence we should not claim any responsibility for them. To do so is, in effect, to denigrate God and his role in our life.

> There is no man living who is not constitutionally liable to falsehood, to theft, to adultery, to murder. . . . There is no person . . . who is not constitutionally liable to gentleness, kindness, patience, generosity, magnanimity, on the presentation of adequate motives. . . . Hence we are not inwardly or spiritually chargeable with this good or evil: they are common to the race of men, and we are consequently forbidden to make any individual approbation of them. . . . All goodness possible to the creature [person] is *ex vi termini* ["by virtue of its function"] a derivation from God to him.[23]

In the context of his father's religious philosophy, it is possible to see why James would have both shunned "moral praise," since it would effectively mean accepting credit for God's will acting through him, and also decreed the affairs of human life, including moral evil, a "phantasmagory": an illusion supported by our assumption of our own autonomy.

By the time he wrote to Ward in 1868, however, James's allegiance to Henry James Sr.'s attitudes had come unstuck. A month earlier, after more than a year of struggling with his back condition, evil had begun to inveigle its way into James's diary and letters. He began to question its presence and essence for himself, from a number of different perspectives, musing on the experience of evil as something done *to* us rather than done *by* us. The first surviving reference is in a diary entry for April 3, in connection with his reading of Homer's *Odyssey*:

> A given evil to the ~~greeks~~ Homeric greeks . . . seems to me to have been thought of as evil only transiently & to those whose lot it was to suffer by it; and they accepted it as part of their inevitable bad luck. Outsiders were not moved to a disinterested hatred ~~& denial~~ of it *in se* [in itself] & denial of its right to darken the world.[24]

In a letter to his brother Henry, he repeated these sentiments, clearly impressed by the "old heathens"; their "cool acceptance" of the universe; and their belief that "the only evil is that every thing in it in turn ceases to exist." He contrasted their confidence with the latter-day conviction that "the world in its very existence is evil," a reference that appears to invoke Christian ideas of "the Fall," when Adam and Eve were expelled from the Garden of Eden, and the suffering and death that are visited on humanity in consequence. For the Greeks there was "no 'reason' behind [evil], as our modern consciousness relentlessly insists." Whereas, in contrast, the "trouble with the modern man wd. be intellectual; he wd. always be trying to get behind Fate, and discover some point of view fm wh. to reconcile his reason to it—either by denying the good of the world,—or inventing a better one on t'other side,—or something else."[25]

James returned to this theme again in his diary, five days later, and was still mulling it over, several weeks later, in his letter to Ward. There he described the contemporary religious determination to "'get around' the contradiction between good and evil, and to fuse them both in some higher unity." We insist, he argued, that

> The naïve natural good of the Greek is not *the* good, but holds a subtle poison—the justness of the just man is not a final fact as it was for the greeks, but, under the name of self-righteousness or what not, merely throws open an endless realm of evil more subtle and strong than that of the senses. And so the evil of the senses is not *the* evil for us.[26]

He concluded that the Greek way of taking life holds an important message: "If we can only bring ourselves to accept evil as an ultimate inscrutable fact," he suggested, "the way may be open towards a great practical reform on earth, as our aims will be clearly defined, and our energies concentrated."[27]

James's newfound interest in evil, and in particular the natural "evil of the senses," might appear, at this stage in the record, to stem merely from intellectual curiosity. But, several months later a personal dimension to his interest in this subject became apparent in his correspondence. In 1869 he wrote to his brother Henry, who was in the midst of one of his acute and persistent attacks of constipation. Sympathizing with his brother's plight James proceeded to lament their respective sickly predicaments from a metaphysical perspective:

> For what purpose we are thus tormented I know not,—I don't see that Father's philosophy explains it any more than any one else's. But as Pascal says . . . there's a divine instinct in us, and at the end of life the good remains, and the evil sinks into darkness. If there is to be evil in the world at all, I don't see why you or I should not be its victims as well as any one else—the trouble is that there should be any.[28]

James's letter suggests that his focus on the Greek attitude toward the misfortune of evil owed something to his own recent experiences of bodily pain and sorrow. This association of evil with pain, and especially the pain of illness and infirmity, is, moreover, a consistent presence in James's correspondence and private notes from this point onward.

In a personal note that is archived with others from these years, James discussed what it means to "hate evil & defy it": "Evil may kill me, but the fear of her can't make me swerve from the pursuit of ends which *in'se* [in themselves] & apart from their consequences I love." He also gave an example of this kind of defiance: "Come, harlot! I'll kiss you because you are a lovable thing, and treat the pox afterwards." The implication, in this hypothetical instance, was that the evil James proposed defying was "the pox," or syphilis. In another example the given evil was the more mundane one of toothache.[29]

In May of 1870 James revisited the topic of "earth's misery" in another letter to his brother Henry. "What new horror is this?" he exclaimed, on hearing of his sibling's latest affliction, a potential kidney problem. He continued his commentary on this horrific development in a philosophical vein:

> It seems to me that all a man has to depend on in this world, is in the last resort, mere brute power of resistance. I can't bring myself as so many men seem able to, to blink the evil out of sight, and gloss it over. It's as real as the good, and if it is denied, good must be denied too. It must be accepted and hated and resisted while there's breath in our bodies.[30]

James linked Henry's ill health with the concept of evil and reiterated his personal stance toward it that he had outlined, two years earlier, in response to the Greek example. It is important to note that James's use of the term "acceptance" did not, in this instance, carry with it a sense of resignation. Instead James appears to have been challenging the theological doctrine associated with Saint Augustine of Hippo, who maintained that evil is simply an absence of good and thus insubstantial in and of itself. In his *Enchiridion on Faith, Hope, and Love*, a treatise on Christian piety, Augustine used the example of wounds and healing to illustrate his argument.[31]

Equally, James's father was also wont to "blink the evil out of sight, and gloss it over." Henry James Sr. dismissed "physical evil" as a necessary, "constitutional fact" of our earthly life as animal creatures, and, as such, "the intellect consequently has no quarrel with [its] existence."[32] In contrast, acceptance for James meant recognizing the evil of the senses as something "real" and "ultimate," not resigning himself to its continuing existence. On the contrary, James proposed that this sort of acceptance is the first step toward melioration; this was fighting talk.

Such an acceptance, however, placed a heavy load of responsibility on James's shoulders. If evil was genuine, a pernicious reality to be contested and fought, then he had a very difficult life choice to make. In a world where evil was real, marriage and children, for an invalid such as himself, would constitute an abdication of his moral responsibilities. This sentiment is expressed clearly in a letter to his brother that he sent at the end of 1869. Bob had announced his intention to marry a woman who suffered, as he did, from the back problems that had tormented many of his siblings. James's medical training left him in no doubt as to the hereditary consequences of such a marriage:

> What results from every marriage is a part of the next generation, and feeling as strongly as I do that the greater part of the whole evil of this wicked world is the result of infirm health, I account it as a true crime against humanity for anyone to run the probable risk of generating unhealthy offspring. For myself I have long since fully determined never to marry with anyone, were she as healthy as the Venus of Milo. . . . I want to feel on my death bed when I look back that whatever evil I was born with I kept to my self, and did so much towards extinguishing it from the world. . . . I would undergo anything myself to escape fm. the guilty feeling of having deliberately put into the world a creature destined to bear such a burden.[33]

James conceded that it would be wrong of him to expect his brother to uphold the same ideals as himself but that, equally, he could not, in good conscience, abstain from making his feelings plain. The letter strongly reinforces the idea that evil, pain, and illness had become very closely related in James's worldview and implies that he felt himself to be personally tainted by their malevolent association with his body. In one note from this period, he discussed both suicide and evil and explained how he felt about his own intimate relations with the latter: "My trouble is that its existence has power to haunt me.—I still cling to the idol of the unspeckled, and when Evil takes permanent body & actually sits down *in* me to stay, I'd rather give up all, than go share with her in Existence."[34]

It is clear from these and other private notes and letters that James's conception of himself as diseased and unfit carried, for him, an ethical as well as a metaphysical significance. He felt his burden of pain to be, in turn, a potential burden on society. There were multiple strands to James's deliberations on this subject. Many of them, during his youth, originated in his espousal of the utilitarian philosophy of John Stuart Mill.

Poisoned with Utilitarian Venom

John Stuart Mill (1806–73) was a mid-nineteenth-century utilitarian who developed and expanded the ideas of his father James Mill (1773–1836) and Jeremy Bentham (1748–1832). Put simply, utilitarianism was a philosophy that preached the doctrine of the "greatest happiness of the greatest number" as the supreme ethical goal.[35] Bentham, the grandfather of utilitarianism, had begun by calling it "the principle of utility" in his earlier writings, but he eventually came to the conclusion that happiness, rather than utility, better captured what he believed should be the driving force behind all of humankind's actions. He intended it as a solution to the problem of subjectivism that he found underpinning most of the theories that prescribed a set of individual moral rights. The principle of utility was a way out of this conundrum, a method of deciding between competing ethical opinions that could provide philosophers and legislators with a reconciling objective standard.[36]

There are several letters, written by James in early adulthood, that confirm his familiarity with the theories of utilitarianism.[37] Such an awareness is not surprising since the 1860s and 1870s were the heyday of these ideas in the popular arena; fierce debate over the issues they invoked regularly raged across the pages of the great Victorian periodicals

in Britain throughout these years, and James was no stranger to transatlantic authors and disputes.[38] In later life he dedicated *Pragmatism* "to
the memory of John Stuart Mill from whom I first learned the pragmatic
openness of mind and whom my fancy likes to picture as our leader were
he alive today." And, in the text itself, he explicitly acknowledged a utilitarian element to his own pragmatic philosophy.[39] What remains unsaid
in these public writings is the extent to which Mill's ideas impacted on
James's personal life.[40]

James's notebooks attest to the fact that he read John Stuart Mill's
Auguste Comte and Positivism at some point during the late 1860s.[41] This
book-length assessment of Auguste Comte's (1798–1857) philosophy
included Mill's own thoughts about the French philosopher's proposal
for a "Religion of Humanity." Although strongly critical of many aspects of Comte's ideas Mill was supportive of the essence of his plan,
which he understood as referring "the obligations of duty, as well as all
sentiments of devotion, to a concrete object at once ideal and real; the
Human Race, conceived as a continuous whole, including the past, the
present, and the future."[42] He claimed that "the Happiness morality" offered "the possibility of giving to the service of humanity, even without
the aid of belief in a Providence, both the psychological power and the
social efficacy of a religion."[43]

From Mill's perspective, one of the chief moral aims should be that
of teaching all citizens "to form the habit, and develop the desire of being useful to others and to the world."[44] This ethical standpoint made
sense to James who was convinced by its practical appeal. In a letter to
his father in late 1867, he conveyed his belief that

> *all* tendencies must now a days unite in Philanthropy; perhaps an
> atheistic tendency more than any, for sympathy is now so much de-
> veloped in the human breast, that misery and undeveloped-ness
> would all the more powerfully call for correction when coupled with
> the thought, that from nowhere else than from us could correction
> possibly come, that we ourselves must be our own providence.[45]

The Religion of Humanity provided James with a purpose and helped
him orient his own life path at a time when ill health had laid waste to
many of his plans. Even if medical practice seemed out of the question,
he could still pursue some sort of research career in "the border ground
of physiology & psychology" that would enable him to leave a useful
legacy, however modest, for humankind. As he explained to his friend
Ward a week later: "I now set my stake in *doing*; anything no matter if no
bigger than a pins head, the discovery of one fact . . . wd. let me sink out

of existence satisfied that I had paid my debt." Such thoughts brought him strength during "the lonesome gloom" of the previous summer, he told him, explaining that he had taken comfort from a "sort of 'Mankind its own God, or Providence' scheme." He confided that "the only feeling that kept me from giving up was that by waiting and living by hook or crook long enough I might make my *nick*, however small a one in the raw stuff the race has got to shape."[46]

James's utilitarian principles may have offered him comfort and direction, but they also exacted a cost: the pressure of results. Mill made it clear that his philanthropic regime relied in part on the association it forged between "self-respect" and the "desire of the respect of others" with "service rendered to Humanity." He also, moreover, explicitly stated the duty that was placed upon a citizen to do everything in their power to maintain their health "since by squandering our health we disable ourselves from rendering to our fellow-creatures the services to which they are entitled."[47] Should James fail to deliver anything that benefited "the welfare of the race," he and his life were worthless, even if illness and infirmity were to blame. James took this message to heart, writing to his brother Henry, during one of his stays at the spa in Teplitz, that his recovery was crucial since it would "turn him from a nondescript loafer, into a respectable working man, with an honourable task before him."[48] From James's perspective, his physical capacity to work was inseparable from matters of respect and honor and, ultimately, his ethical philosophy. However, this focus on practical achievement was to become, as time went on, untenable to James. When illness made all activity and even the future possibility of useful work seem impossible, his utilitarian ideals came under fire.

This disenchantment was evident in a letter to Ward in the autumn of 1868. Writing in a "dilapidated state" of health and spirits, James complained of the "inanity" of his recent activity and of how little fitted his physical frame was to his chosen career in science. Since April that year he had been voicing doubts about his ability to pursue physiological study, on the grounds that his ongoing back issues would make laboratory work permanently impossible. The implications of not being able to make any meaningful contribution had clearly been preying on James's mind. He confessed to his friend that he was "poisoned with Utilitarian venom, and sometimes when I despair of ever doing anything, say: 'Why not step out into the green darkness?'" It is clear from the passages that follow that James had been considering whether his life was one worthy of continuation. In a letter to his friend Holmes a few months earlier, he had introduced the question that "all men who pretend to know themselves ought to be able to answer," namely: "'what reason can you give for continuing to live? What ground allege why the thread of your days shd. not be snapped *now*?'" In answering for himself he acknowledged ominously that "I could wish that my general grounds were more defined than they are."[49]

James's personal papers indicate that this question became even more pressing over the next year. In a collection of notes that appear to date from 1869, he evaluated whether a utilitarian perspective should direct him to suicide as being the best solution for the common good.[50] The notes contain a detailed analysis of the interrelations between pain, pleasure, sympathy, and suicide. The first two concepts of "pleasure" and "pain" had always been fundamental to utilitarian philosophy ever since Bentham declared that "nature has placed mankind under the governance of two sovereign masters, pain and pleasure." This assumption, of the intrinsic primacy of these two values, was the starting point of his rule of utility.[51] He intended that lawmakers should consider the resulting aggregate of society's pleasures and pains, or its collective happiness, when choosing a course of action.[52] Similarly, James had adopted this governmental directive as a guide and inspiration for his personal life decisions. In his own words, he had taken consolation in the thought that he belonged "to a brotherhood of men possessed of a capacity for pleasure & pain of different kinds" whose brotherly "enjoyment" or welfare was a worthy ideal.[53]

The concept of "sympathy," meanwhile, was the subject of much cultural reflection during the late nineteenth century, and in the ethical literature it was implicated in the adoption of altruistic actions and social evolution. It also carried a different meaning than it does today. For James and his contemporaries, it was a more emotionally involved process of identification that encompassed a sense of actively taking on someone else's suffering, as opposed to merely acknowledging its validity.[54] James, for example, wrote to his father about keeping the details of his back condition secret from various friends and relatives, on the grounds it would cause them "useless pain," and insisted that "poor Harry especially, who evidently from his letters runs much into that utterly useless emotion, sympathy, with me, had better remain ignorant."[55]

In the notes mentioned above, James put it bluntly: "sympathy gives pain," and then went on to ask "Shd. sympathy go so far as to dictate suicide? As when I sick, become but an eyesore & stumbling block to others." He proceeded to discuss the ideal way to deal with "evil," observing that "in a given case of evil the mind seesaws between the effort to improve it away, and resignation." Further on, he clarified that "resignation affords ground and leisure to advance to new philanthropic action. Resignation should not say, 'It is good,' 'a mild yoke,' and so forth, but

'I'm willing to stand it for the present.'" He concluded that, with respect to his initial question about suicide, there were "three quantities to determine. (1) how much own pain I'll stand; (2) how much other's pain I'll inflict (by existing); (3) how much other's pain I'll 'accept,' without ceasing to take pleasure in their existence."[56]

There are many instructive elements to James's deliberations. His discussion of "evil," and how best to approach it, is consistent with the supposition that, by this point in time, he had come to identify evil with illness and pain. His final analysis exemplifies a utilitarian calculus of happiness, weighing up the total pains and pleasures accrued to himself and society by the fact of his existence. First and foremost, however, these notes indicate how much intellectual energy James spent reflecting on whether he should end his life and the multifarious ways these ruminations connected with his developing philosophical ideas.

The Ethics of Self-Destruction

In the archives at Harvard are a series of notebooks kept by James over his lifetime. Number 26 is different from all the others, however. It is an *index rerum*: a book of ruled pages that are bound in a tabulated and alphabetized format for the systematic preservation of quotations and references. The notebook is dated, in James's hand, November 25, 1864, which presumably marks the earliest date of the entries inside. And, according to the archivists, the later contributions were written up until and including the year of 1890. In the margin of each page is a list of keywords, entered by the notebook's author, which denote the subject of the citation or reference next to them. These keywords represent, in some form, an indication of James's prevailing interests during this period of his life. While some topics are mentioned fleetingly, others he returns to many times, and suicide is one of them.[57]

Distributed among allusions to, and facts about, the "sympathetic nervous system," "skulls," "superstition," and "supersaturated solutions" are four separate entries listed under "suicide." Two of the four entries, moreover, consist of multiple references, and altogether James lists a total of nineteen different authors' writings on the matter. It was a theme to which James appears to have returned time and again in his studies, one that held an abiding fascination for him. The indexed references in question are varied and comprehensive, drawn as they are from the fields of philosophy, literature, psychiatry, and poetry and ranging in date from the seventeenth to the nineteenth centuries.

The majority of these works focus on the linked questions of whether suicide is justifiable or desirable and on what terms. The French philosopher Madame de Staël's *Réflexions sur le suicide*;[58] Jean Dumas's *Traité du suicide*;[59] a Latin treatise on the subject by the Swedish philosopher Johannes Robeck;[60] John Donne's *Biathanatos*;[61] David Hume's essay on suicide;[62] a poem of Samuel Taylor Coleridge's entitled "The Suicide's Argument"; and passages in Michel de Montaigne's *Essays*;[63] Jean-Jacques Rousseau's *La nouvelle Héloïse*;[64] and Thomas De Quincey's *Note Book of an English Opium-Eater* are all part of this centuries-long debate.[65] Some, such as Donne and de Staël, approach the question from an ostensibly Christian perspective; others invoke variations on the ancient arguments of the "pagan" philosophers.

The Stoic arguments in favor of suicide were rehearsed by Montaigne, Hume, Voltaire, and Montesquieu and described in a nineteenth-century history text, William Lecky's *History of European Morals*, all of which James referenced in his index in connection with this subject. In addition, James read and admired the *Meditations* of Marcus Aurelius as a young man, together with the writings of Seneca and Epictetus, and there is considerable evidence that he subsequently adopted a philosophical worldview that was inherently Stoic in many respects.[66] All of this goes to suggest that Stoicism comprises an important intellectual context within which to locate James's contemplations on suicide. It is also clear from a letter written by one of his philosophical associates that he and James discussed, on more than one occasion, the rationality or otherwise of suicide according to Epicurean principles.[67] Epicureanism, meanwhile, was closely associated with utilitarianism in the minds of James and his contemporaries, as both philosophies shared a hedonistic focus on the pursuit of pleasure and the avoidance of pain as the primary ethical "good."

In contrast to the hostile "monolithic stance of Christianity," in antiquity the attitude toward suicide varied from one philosophical school to another, and this diversity was reflected in the laws enforced across the different cities.[68] Where Plato's writings were ambiguously prohibitive, and Pythagoras's militantly so, the Epicurean and Stoic authors were openly approving. The enduring popularity of Stoicism, coupled with the permissive imperial Roman law, are credited with what has been historicized as an unusual proclivity for self-murder among the Roman elites. Writers from the sixteenth to the eighteenth century have made reference to the famous Roman suicides of Cato, Cassius, Brutus, and Seneca.[69]

Epicurean and Stoic philosophers justified suicide as "a legitimate way to end a life of unremitting pain" or as "an act of nobility, and expression of political or philosophical principle."[70] These political suicides assume a significant presence in the accounts of Roman historians, throughout the ages, who report them as heroic acts sanctified by a contemporary philosophical belief in the supreme worth of the individual, "whose liberty resided in the ability to choose whether to live or die."[71] Cato, for example, is commended in countless early modern and modern accounts for dying by his own hand. His death is glorified as the ultimate protest against the tyranny that marked the end of the Roman republic and as the honorable act of a man willing to die for his principles.[72]

A Stoic suicide need not be glorious or born of political persuasions, however. Old age was, in and of itself, sufficient justification. Seneca (ca. 4BCE–CE65) insisted that, while there is no reason to kill oneself as long as the body and mind enjoy full possession of their faculties, there is no compulsion to continue to live with the decrepitude of later life: "If, on the other hand, the body's past its duties, it may be (why not?) the right thing to extricate the suffering spirit." He clarified that such an exit should not be in order to escape pain, which would be "weak and craven," but because the "crumbling, tumble-down tenement" of a body is "likely to prove a bar to everything that makes life worthwhile."[73]

Similarly, any circumstances that make it impossible to live with purposeful intent, including ill health, were reason enough for a Stoic suicide. The breadth of these grounds helped make suicide seem banal among the wealthy classes of imperial Rome.[74] The Stoic philosopher Epictetus (ca. 55–135) pronounced, for example, "What does it matter by what road you enter Hades?" and his successor Marcus Aurelius (121–180) wrote: "As you intend to live when you depart, so are you able to live in this world; but if they do not allow you to do so, then depart this life, yet so as if you suffered no evil fate. The chimney smokes and I leave the room. Why do you think it a great matter?"[75]

James's personal letters attest to the fact that, during the painful years of his young adulthood, Stoicism offered him a philosophy of consolation and resolve in the face of suffering and disappointed aspirations. It conferred a sense that the universe was unfolding as it should, even if this underlying "Reason" was not fathomable from his own limited viewpoint. In James's words to his regular confidant Ward, "The stoic feeling of being a sentinel obeying orders without knowing the General's plan is a noble one."[76] However, his reading lists reveal that James was only too aware that as his physical state deteriorated further, his Stoic heroes would have had no qualms about choosing death over life. Whether he approached his dilemma from the perspective of Stoicism or utilitarianism, there was no compelling reason for him to continue to endure an unproductive existence dogged by unrelenting pain. On the contrary, it was conceivable that death might be the more socially responsible or honorable course of action. But these schools of philosophy were not the only context for James's study of suicide.

Alongside the ethical discussions of suicide listed in James's *index rerum* are references to the works of several medical authors.[77] For these authors, whose legacy is culturally dominant today, suicidal inclinations were to be viewed, above all else, as a symptom of illness. One such British physician noted by James, Forbes Winslow (1810–74), expressed his belief, in no uncertain terms, that "the disposition to commit self-destruction is, to a great extent, amenable to those principles which regulate our treatment of ordinary disease; and that, to a degree more than is generally supposed, it originates in derangement of the brain and abdominal viscera."[78]

In 1840, when Winslow published this opinion stating that suicide was in many cases a symptom of disease, the act was still a criminal offense in England. It was only via a verdict of insanity that the family could clear their relative's name posthumously and retain the proceeds of the estate of the deceased. In these circumstances the diagnosis of self-destruction was a fiercely contested activity. In many forums the underlying question of who could justly be held responsible for their own actions came to be linked with the novel, and controversial, medical category of "moral insanity." Suicide was just one of many legally problematic behaviors that could be explained, according to some doctors, by a diagnosis of this sort.

The long-standing legal definition of insanity in Britain, enshrined in the writings of the eighteenth-century Lord Chief Justice Matthew Hale, was couched in terms of a total, albeit sometimes temporary, loss of reason.[79] During the early decades of the nineteenth century, however, medico-legal negotiations expanded the boundaries of insanity. An impairment of the intellect was no longer required, and instead those suffering from a derangement in their volitional or emotional "moral" functioning could also qualify. This disputed concept of "moral insanity" was often allied with another new diagnosis, "monomania" (a madness exhibited only in connection with one topic, or faculty), in the writings of contemporary commentators. In the nineteenth-century courtroom, a defendant who knew their actions to be wrong but was powerless to stop themself could now attempt to claim insanity: "Traditionally common law had ascribed a 'wicked will' to wicked persons. . . . But in the nineteenth century no longer are evil thoughts one's own; they are the residues of *diseased* emotions or morbid pathology of the will itself."[80]

Many observers were concerned by this trend including the anonymous reviewer of another publication on suicide, which was cited by James in his index, and written by the French catholic doctor Louis

Bertrand.[81] The reviewer, writing for the *American Journal of Insanity* in 1857, opined that he was "glad, however to see in the prominence given to the moral relations of suicide by Dr. Bertrand, as by most others who have recently written upon the subject, a blow at the theories of monomania and moral insanity which have been extensively urged."[82]

The reviewer's approval of Bertrand's insistence that suicide should be condemned as a sin against God, one's family, and one's country, betrayed a very different viewpoint than that held by Winslow, who was an avid supporter of the concept of monomania.[83] Winslow's insistence on the physical origins of the suicidal impulse was in keeping with his general stance on the medico-legal relations of insanity. He wrote several controversial papers on this subject and, as one of the most extreme advocates of the nonresponsibility of the insane, was instrumental in establishing medical grounds for the plea of insanity in court cases.[84] Along with this narrowing of the field of criminal responsibility, however, the new diagnoses brought with them a broadening of the territory of the lunatic. In this sense the early nineteenth-century construction of "moral insanity," and its sibling "monomania," brought madness much closer to the minds of many.

Several contemporary authors have drawn attention to a fear of insanity said to have haunted James. They cite his account of a terrifying personal vision when he conceived of his potential transformation into an "epileptic patient whom I had seen in the asylum." He sat there, wrote James, "with greenish skin, entirely idiotic . . . moving nothing but his black eyes and looking absolutely non-human. . . . *That shape am I*, I felt, potentially."[85] Reading Winslow's textbooks on insanity it is not hard to understand the grounds for such fears. James wrote to his cousin, Katharine (Kitty) James Prince, whose husband ran the newly opened Northampton Lunatic Hospital, shortly after reading one of Winslow's books for the first time, aged twenty-one. He had previously been considering following Dr. Prince into a career in asylum medicine but had subsequently come to the conclusion that

> as for "live lunatics" I am very much afraid that I am little fitted by
> nature to do them any good, . . . I verily believe I shd catch their
> contagion & go as mad as any of them in a week, for on reading Dr.
> Winslow on Obscure Diseases of the Mind not long since, my rea-
> son almost fled, it was so rudely shaken by the familiar symptoms the
> Doctor gave of insanity.[86]

It is hard to determine from their tone how seriously James intended his comments to be read, but he valued Winslow's book enough to

1.3 James's personal copy of Forbes Winslow's *On Obscure Diseases of the Brain, Disorders of the Mind: Their Incipient Symptoms, Pathology, Diagnosis, Treatment, and Prophylaxis* contains a number of bookmarked pages, including this one on the subtle indications of suicidal insanity and the internal experiences of those tormented by thoughts of self-destruction. The poem at the bottom of the page reads: "He hears a voice WE cannot hear, Which says, HE must not stay, He sees a hand WE cannot see, Which beckons HIM away." (Winslow, *On Obscure Diseases of the Brain*, 248–49 [Houghton Library, Harvard University].)

keep it. A copy of the 1860 edition, with James's nameplate affixed to the inside cover, is housed in the Harvard Library's collection. It would appear to be the original copy that inspired his remarks, and it is clear from the presence of marginalia, typical of the style he used to mark up books later on in life during his lecturing career, that it was a work that he referred to on occasion.[87] There are also four small scraps of paper, inserted as bookmarks, tucked tightly into the binding between particular pages. Although it is impossible to confirm that these were put there by James's hand, they are made of old, acidic paper that has stained the pages between which they have lain sandwiched, it would appear, for many years. In the context of James's fear of insanity, his suicidal contemplations and his long-term problems with insomnia, the pages bookmarked are suggestive and refer to sections on the "symptoms of latent insanity"; "subtle types of suicidal insanity"; "insanity commencing in a dream"; "general principles of treatment and prophylaxis."[88]

The overall thesis of Winslow's book was that cases of serious "brain disorder," such as "apoplexy, acute softening, paralysis, epilepsy, meningitis, cerebritis, or mania," may be detected at an early stage. Winslow went on to warn the reader though, that these early manifestations of brain disorder were of "so slight and transient a character" that they were "easily overlooked by the patient, as well as by his physician." He proceeded to list some examples, many of which were present in James's reports home on his own condition: "a condition of general, or local muscular weakness . . . a state of ennui . . . impairment and disorder of the sense of sight . . . an inability to concentrate the attention continuously on any subject." According to Winslow, they were *all characteristic symptoms, frequently diagnostic of disease having commenced in the brain.*"[89]

James also alluded in correspondence to his "low spirits" and "melancholy" feelings.[90] In the context of the medical theories of the time such low moods were taken extremely seriously: they were portents of the utmost significance. In one chapter in his book Winslow had this to say about periods of despondency:

> There are few morbid mental conditions so fatal in their results as these apparently trifling, evanescent, and occasionally fugitive attacks of depression. They almost invariably (in certain temperaments) are associated with a disposition to self-destruction.[91]

According to Winslow, and other medical authorities of the era, James's symptoms and his self-confessed despondency potentially indicated a serious cerebral pathology. They presaged the likely onset of severe brain disease and even, ironically given James's deliberations as to whether or not his life was worth living, uncontrollable suicidal impulses. Winslow referred to cases where unfortunate individuals came to believe themselves pursued by "unseen demons" impelling them to self-destruction. He also recounted other instances where those who had previously experienced a "lowness of spirits" went on to develop apoplexy or die from "softening of the brain."[92] Worse still, this physicalist model of mental ill health could promise little hope of cure. All James could do was wait and worry; in time his "cerebral mass" would pass its sentence on his fate.[93]

Henry Maudsley (1835–1918), another renowned English alienist whose work was familiar to James, aptly summed up the frightening implications of such psycho-physical doctrine. His writings referred to the inevitable and unchallengeable authority of man's physiological structure, and in particular his brain, as the "tyranny of his organization." Maudsley proclaimed that "in consequence of evil ancestral influences, individuals are born with such a flaw or warp of Nature that all the care in the world will not prevent them from becoming . . . insane. . . . No one can escape the tyranny of his organization; no one can elude the destiny that is innate in him."[94] There are indications, moreover, that James found himself for a time increasingly, albeit reluctantly, convinced by the philosophical paradigm that underpinned this kind of medical creed.

Conscious Automata

In 1869 James wrote to his friend Ward declaring how, in his own words, he felt "swamped in an empirical philosophy—I feel that we are Nature through and through, that we are *wholly* conditioned, that not a wiggle of our will happens save as the result of physical laws."[95] James's son, in his edition of his father's correspondence, footnoted James's remarks with the following comment: "It ought perhaps to be noted," wrote Henry James III, "that at about this time a man whose philosophic ability was great and whose thought was vigorously materialistic was often at the house in Quincy Street. This was Chauncey Wright."[96]

Chauncey Wright (1830–75) was a close friend of the James family from as early as 1864. He was a passionate philosopher whose incisive intellectual prowess was commented on by many who knew him. During the early to mid-1860s and around the time he first came into contact with James, Wright's own philosophical position went through a fundamental shift, and he became wholly converted to the empiricism of John Stuart Mill and the evolutionary naturalism of Charles Darwin.[97] His intellectual contributions have largely been forgotten, overshadowed by those of his better-known sparring partners such as Charles Sanders Peirce, Oliver Wendell Holmes Jr., and James himself, but he did publish a handful of papers before his untimely death at forty-five.[98]

In these articles many of Wright's fiercely maintained views can be discerned, including his opinion on the interrelations of mind and body. On this topic he had the following to say:

> The conditions of the nervous tissues . . . are a sum of material conditions, which, occurring along with other material conditions around them determine particular perceptions, thoughts, and volitions as

mental events . . . mental events and their combinations are fully conditioned, as events, on material ones.[99]

That James found this position disturbing is implicit not just in the letter
to his friend Ward but also in his tribute to Wright that was published
in the *Nation* shortly after his death. There James remarked that "of the
two motives to which philosophic systems owe their being, the craving
for consistency or unity in thought, and the desire for a solid outward
warrant for our emotional ends, his mind was dominated only by the
former."[100] James's notebooks and an unpublished essay from this period also testify to his persistent and intense intellectual wrestling with
"Empiricism," which was how he characterized Wright's philosophical
outlook. Tellingly, he concluded the essay, on brain processes and feelings, with a discussion of the implications of "the consistent empiri-
cist's" position for the diagnosis of insanity.[101]

James's concerns about his own mental condition appear to have
been one of the major issues at stake in these wranglings with the empiricism of his contemporaries and the deterministic relationship it entailed between neurological activity and thought. His reading lists and
the book reviews that he published during the early 1870s all suggest that
James's focus on mind-body relations during these years was oriented
around mental pathology and its prevention and treatment.[102] These included a review of Isaac Ray's *Contributions to Mental Pathology*; a commentary on the section concerning "morbid states" in William Carpenter's *Principles of Mental Physiology*; and an essay on "Mental Hygiene"
that referred to the work above and also to Maudsley's *Responsibility in
Mental Disease*.[103]

A letter sent by James's father to his brother Henry in March of 1873
corroborates, retrospectively, this medical focus and its personal significance to James. In this correspondence Henry James Sr. recounts how
James had come into the room, a few days earlier, and remarked on how
well he had been feeling, mentally. In comparison with his condition
a year ago he felt his mind to be "so cleared up and restored to sanity"
that it was "the difference between death and life." Apparently, when
asked by his father what had brought about this change in his spirits,
he had replied:

> The reading of Renouvier (specially his vindication of the freedom
> of the will) & Wordsworth whom he had been feeding upon now for
> a good while; but especially his having given up the notion that all
> mental disorder required to have a physical basis. This had become
> perfectly untrue to him. He saw that the mind did act irrespectively

of material coercion, and could be dealt with therefore at first hand, and this was health to his bones.[104]

There is no clue in the letter as to what theoretical form James's new understanding of the mind and body took. A year later, however, he wrote a review of William Carpenter's *Principles of Mental Physiology* that provides a window onto the drift of his ideas at this time. Once again, the concept of pain was critical to his thinking.

In his Carpenter review, James took on two of the intellectual giants from the contemporary field of psychology: Herbert Spencer and Alexander Bain (1818–1903). He maintained that both authors described consciousness as a "chain" of mental events associated with, but unable to influence, the chain of physical events taking place in the brain and nerves; "just as a shadow accompanies a man walking, but does not influence his gait." To dispute this assumption that the line of influence between body and mind travels only in one direction, James introduced "that very general law of pleasures and pains which associates the former with occurrences that further the well-being of the animal, the latter with influences which are deleterious." Spencer and Bain both put this law to work in their theories of psychological development, James argued, but fail to spot how "subversive" it is of their doctrine of mind-body relations. He cited the specific example of arrested breathing, which is accompanied by "agonizing distress." He pointed out how it "*seems* as if the struggles made to get air were a direct consequence of the distress felt" and questioned whether the same "violent movements to procure oxygen, and so forth" would occur if the mental response experienced in this situation was one of pleasure.[105] In another book review the following year, he stated his case more succinctly: "Taking a purely naturalistic view of the matter, it seems reasonable to suppose that, unless consciousness served some useful purpose, it would not have been superadded to life."[106]

In an early set of public lectures, which he gave in 1878, he examined this argument in more detail. The manuscript notes of these Lowell Lectures, on "The Brain and the Mind," reveal that James argued for the influence of the mind, as opposed to it being a mere epiphenomenon of the brain and human beings simply "conscious automata." According to such theories, James reasoned, we are "victims of delusion when we think that our thoughts produce each other, Good news joy . . . disappointed hopes, sadness." But if consciousness is impotent and "feeling" is "inert, uninfluential, a passenger," James asked, why do we have them? Surely, he insisted, the most likely explanation is that consciousness is useful to the human organism, that it, in some way, "loads the dice" in favor of our survival.[107] In short, he argued, it is legitimate to suppose that our thoughts may direct other thoughts, and ultimately our bodies, rather than assume, as the materialists did, that the opposite was exclusively true. He subsequently published his discussion of this topic in the journal *Mind*, under the title "Are We Automata?" In a letter to fellow philosopher Shadworth Hollway Hodgson (1832–1912), he confided that he had written this article "against the swaggering dogmatism of certain medical materials, good friends of mine, here and abroad. I wanted to show them how many empirical facts they had overlooked."[108]

According to James's father, however, there was another, secondary strand to James's philosophical development during this period. His letter also mentioned James's enthusiasm for the work of the French philosopher Charles Renouvier (1815–1903) and the case he made for the doctrine of "free will."[109] James's official biographer Perry made this issue central to James's early adulthood. He denoted these years of his life as a period of "soul sickness." During this time, he asserted, James was miserable because he was struggling to find a philosophical rationale for the concept of "free will."[110] Subsequent interpretations have developed this theme, and expanded on the crucial role that Renouvier played in James's solution.[111] What no account has elucidated to date is the part that James's perception of himself as an invalid and his preoccupation with the evils of pain and illness appear to have played in his attitude toward nineteenth-century debates about free will.

For many of James's contemporaries, including those involved in the medico-legal disputes over the definition of insanity and the justifiability of insanity pleas, the free will controversy circled around matters of responsibility and blame. The question on which it turned was the extent to which we are free agents, at liberty to choose between good and wicked actions, and thus accountable for our crimes, small and large. James, in contrast, understood the primary ethical significance of determinism to lie elsewhere, as a letter to a friend and colleague, Francis Herbert Bradley (1846–1924), reveals:

> As for the connection of chance or non-chance action with "responsibility," "merit," and other ethical categories, . . . it has always seemed to me . . . that this whole aspect of ethics is relatively unimportant. The *deep* questions which the moral life suggest are metaphysical: 1) Is evil *real*? 2) Is it *essential* to the Universe? If these questions are to be answered with a yes I care not who is responsible, or may be called so.[112]

For James determinism and ethics were both mired in the same issue about the meaning and significance of evil. In his 1884 essay "The Dilemma of Determinism," he explained that, for him, determinism represented the conviction that the universe is unambiguous in every way, and everything we find within it, including all our experiences of evil, are meant to be and could never be otherwise. Evil, in other words, is inevitable. We have no hope of avoiding it, and, moreover, it makes no logical sense to feel "regret" about that fact since no other outcome was ever possible. James, on the other hand, defended an indeterministic alternative, namely, the "free-will theory of popular sense based on the judgement of regret," which represented a world that is "vulnerable, and liable to be injured by certain of its parts if they act wrong. And it represents their acting wrong as a matter of possibility or accident, neither inevitable nor yet to be infallibly warded off." In short, he equated "free-will" or indeterminism with chance: "the chance that in moral respects the future may be other and better than the past has been." Or, in other words, James understood the doctrine of free will to represent the possibility that the future may hold less evil than the present.[113]

This construal of the free will question held an explicitly medical significance for James and perhaps others too. In the "Dilemma of Determinism" he referenced a recent gruesome crime as an irrefutable instance of evil; the perpetrator in question had brutally murdered his wife. At first sight this example appears unconnected to James's tendency to associate evil with illness. As discussed earlier, however, in the psychiatric and sociological literature of the day, such criminality was frequently associated with a diagnosis of mental disease and physiological factors, and in earlier book reviews, James had made his own sympathies with this perspective clear.[114] A surviving item of correspondence between James and the English author and proponent of evolutionary ethics Leslie Stephen (1832–1904) confirms this interpretation. Stephen wrote to James, continuing what appears to have been a discussion between them that had begun at an earlier date:

> As for your murderer, I dont see why you should "regret" the fact more than I, for what is past is necessary. But if by regret you only mean that the fact disgusts you & that you will try to suppress similar facts in future I have, so far as I can see, just the same right to regret as you. Indeed but for my belief in "determinism," I should be the less entitled to hope that they can be suppressed; if diseases don't obey laws, why should we trust to medicine?[115]

There is no letter that indicates how James answered Stephen's question, but the implication is that they both located this determinism debate within a medical framework. Set in the biographical context of James's

back pain and mental ill health, it would seem that free will, as a philosophy of indeterminism, represented, for him, the chance that his own
situation might improve. It was the intellectual foundation on which
hope and resilience were made possible; the hope that eventually, perhaps, his future and others' would be less blighted with the evils of illness and pain.[116]

A diary entry for April 30, 1870, recorded James's adoption of this
new perspective and how he applied it to his own life. He wrote of how,
"hitherto, when I have felt like taking a free initiative, like daring to act
originally, ... suicide seemed the ~~only~~ most manly form to put my daring
into." He then proposed a new way of thinking about his future, however,
one that was predicated on a belief in the reality of his own "free will"
and "creative power," a life built on "doing and creating and suffering."[117]

James's championing of his own moral autonomy offered something
more than the potential for recovery and a faith in his own creative stake
in his future, however. His belief in the doctrine of free will was the basis on which he was able to construct a more meaningful life regardless
of his state of health, one that afforded the opportunity for self-respect,
even for an unproductive invalid. Whereas Mill's utilitarian ethics explicitly connected tangible contributions to society with ethical merit,
James's correspondence suggests that he had begun tentatively exploring the possibility and potential of a different kind of moral life, one that
celebrated effort over achievement. In an earlier letter to his friend Ward,
where he bemoaned his own utilitarian beliefs, he wrote of how he en-
vied his friend who traveled on his "free-will and moral-responsibility."
He explained that for those like Ward: "no step they make is trivial. ...
To get at something absolute without going out of your own skin! To
measure yourself by what you strive for & not by what you reach! They
are a superb form of animal."[118]

It becomes apparent, in subsequent letters, that James began to re-
orient his own attitude along these lines. He wrote to his brother Henry,
six months later, around the time of the diary entry above, discussing
the intransigent nature of his back condition: "Strange to say, I feel quite
indifferent to the damned thing—and have (any how for a time) cast
off that slavish clinging to the hope of *doing* s'thing wh. has been the
torment of my life hitherto, with a mental exhilaration I have been devoid of for a long time." Five or so weeks later he wrote again, reiterating and expounding his new position. He explained that his back was
slightly better, and he held out some hope that it may improve further
over the summer months, but his health was now a matter of less concern to him, he insisted, "for I have loosed the lockjaw grasp with wh.
I clung to the hope of accomplishing external work, and transferred my

interest in the game of life to the subjective attitude, *i.e.* become moral-
ized, in some sort."[119]

According to this kind of ethical perspective, James didn't have to
produce any world-changing scientific discoveries to be a respectable
member of society. A commitment to continue fighting his own evil,
rather than capitulate to it, was all that was required. He explained this
line of thinking to his brother Bob, two months later. There he brought
up the subject of his brother's back condition and the constraints it
placed on his life and regretted his inability to console him with religion or philosophy:

> Why you should have been picked out to swell the billions on whom
> fate has laid her rough hand, who can say? But one thing is certain,
> that . . . we learn of resources within us, of whose existence we should
> else have remained ignorant; of power to resist pain . . . and generally
> to keep our heads up where nothing but pure courage will suffice.[120]

There is comfort to be found, his letter continued, in the "inner solitary
room of communion with [your] own good will," namely, the knowledge that you have kept up a "true and courageous spirit." He ended with
what appears to be a quotation from the writings of the Stoic philosopher Seneca on providence: "A strong man battling with misfortune is
a spectacle for the Gods" and, added James, reversing the meaning, so
"is a weak man or any man."[121]

Over the years that followed, James's private labors with evil, and the
ethics of suicide began to surface in his public lectures and articles. In
the manuscript notes that correspond to his Lowell Lectures on "The
Brain and the Mind" of the 1870s, this personal context is brief and elu-
sive. There James explained that the moral life consists of choices between conflicting interests and went on to give examples, the last of
which was whether "to escape pain or at any cost of pain to do some
service to humanity."[122] In a journal article entitled "Rationality, Activity and Faith," the autobiographical subtext to his ideas is vivid and unmistakable. Pain stalks the narrator throughout.

This article, written in 1880 and published in the *Princeton Review* in
1882, brings together elements of many of James's better-known texts
that he produced during the same period and those that followed in the
years to come, including "Reflex Action and Theism" (1881), "The Moral
Philosopher and the Moral Life" (1891), "Is Life Worth Living?" (1895),
"What Makes a Life Significant" (1899), and, most notably, "The Will
to Believe" (1896) and "The Sentiment of Rationality" (1897). The last

contains material duplicated from "Rationality, Activity and Faith," and both the latter two map the central thrust of the essay, which comprises an argument in favor of religious or moral creeds.

James's discussion was aimed at contemporaries such as the English biologist Thomas Huxley (1825–95) and mathematician William Clifford (1845–79) who asserted that religious faiths in the absence of "scientific evidence" were intellectually and morally indefensible.[123] James's counterargument was that, in choosing between two logically consistent belief systems, the more "rational" choice was the one that may "awaken the active impulses or satisfy other aesthetic demands far better than the other." James maintained that the "ultimate principle" of a philosophy "must not be one that essentially baffles and disappoints our dearest desires and most cherished powers." It is for this reason, he argued, that pessimism is unsuited to most men. "Witness the attempts to overcome the 'problem of evil,' the 'mystery of pain,'" he wrote, implying that "evil" and "pain" were synonymous. There is, he added, "no 'problem of good.'"[124]

James insisted that faith was as much at work in the fields of scientific endeavor as it was in religious practice and explained that it was an indispensable part of life: "We cannot live or think at all without some degree of faith. Faith is synonymous with working hypothesis." According to James, we all act from positions of faith, and over time these various beliefs are corroborated or extinguished by the outcomes of our experiments in life. Moreover, he continued, there is "a certain class of truths of whose reality belief is a factor as well as a confessor." These truths, "cannot become true till our faith has made them so."[125]

Further on in the essay James took an example of this kind of truth, namely, that "every human being must some time decide for himself whether life is worth living."[126] He then followed through the two different responses available to someone facing this question:

> Suppose that in looking at the world and seeing how full it is of misery, old age, of wickedness and pain, and how unsafe is his own future, he yields to the pessimistic conclusion, cultivates disgust and dread, ceases striving, and finally commits suicide. He thus . . . makes of the whole an utterly black picture illumined by no gleam of good. . . . The man's belief supplied all that was lacking to make it so, and now that it is made so the belief was right.[127]

Alternatively, James continued, suppose with the same "evil facts," the man makes the opposite choice:

Suppose that instead of giving way to the evil he braves it and finds
a sterner, more wonderful joy than any passive pleasure can yield in
triumphing over pain and defying fear. Suppose he does this successfully, and however thickly evils crowd upon him proves his daunt-
less subjectivity to be more than their match. . . . Will not every one
instantly declare a world fitted only for fair-weather human beings
susceptible of every passive enjoyment, but without independence,
courage or fortitude, to be from a moral point of view incommensu-
rably inferior to a world framed to elicit from the man every form of
triumphant endurance and conquering moral energy?[128]

In this passage James implied that he believed that a certain degree
of evil and pain were a justifiable element of a moral universe. This inference is confirmed by his subsequent introduction of the writings of
James Hinton (1822–75). Hinton was an English surgeon and author of
several popular texts, on medical and ethical topics of the day, that were
widely read in America as well as Britain. In the footnotes, James noted
that his 1866 book *The Mystery of Pain: A Book for the Sorrowful* "will
undoubtedly always remain the classical utterance on this subject."[129]
James cited Hinton's insistence that "inconveniences, exertions,
pains: these are the only things in which we rightly feel our life at all. If
these be not there, existence becomes worthless, or worse." In the rest
of the quotation, Hinton alleged that human beings are made to find
"enjoyment in endurance." When our pains become "unendurable, awful, overwhelming, crushing, not to be borne save in misery and dumb
impatience, which utter exhaustion alone makes patient" this meant, according to Hinton, not that our pains are too great, but that "*we are sick.
We have not got our proper life.*" In this light, he concluded, "pain is not
more necessarily an evil, but an essential element of the highest good."
At this point, James picked up the refrain himself and turned the discussion back to faith, adding that the "highest good can only be achieved
by our getting our proper life, and that can only come about by help of
a moral energy born of the faith that in some way or other we shall succeed in getting it if we try pertinaciously enough."[130]
Sickness and health were stitched into the fabric of James's moral
universe, a universe in which pain's existence must be accounted for but
eventually, via strenuous personal effort, overcome. The religious faith
that James outlined in this essay, and others from this time, comprised
a delicate balancing act. It was predicated, in part, on his need for a gov-
erning Providence, the certainty and assurance that, ultimately, the universe was in safe hands. In the "Dilemma of Determinism" he phrased
this as "an ultimate peace behind all the tempests, . . . a blue zenith above

all clouds."[131] Equally, however, it was a faith that placed more emphasis on human action than divine intervention. In James's words: "This world *is* good, we must say, since it is what we make it, and we shall make it good."[132] The ethical world that he depicted demanded the active participation of humankind to bring about its final, happy, healthy outcome, while at the same time providing sufficient reserves of hope and resilience to sustain them in the battle against evil.

In contrast, throughout "Rationality, Activity and Faith," James implicitly allied the materialism of his scientific colleagues with "fatalism" and the absence of moral motivation. Once again it was *medical* materialism that James had in his sights. Borrowing from his physician's lexicon, he allied the philosophy of materialism with the practice of anesthesia. A philosophy such as this, he argued, fails to galvanize us into action and sanctions instead a shallow Epicurean pursuit of pleasure and the avoidance of pain at the expense of all else.

> *Anaesthesia* is the watch-word of the materialist philosopher brought to bay and put to his trumps. *Energy* is that of the moralist. Act on my creed, cries the latter, and the results of your action will prove the creed true and that the nature of things is earnest infinitely. Act on mine, says the epicurean, and the results will prove that seriousness is but a superficial glaze upon a world of fundamentally trivial import.[133]

At the core of James's early thinking, the center of gravity around which everything else revolved, was the concept of pain. It structured the narrative of his days and the long dark night of his soul. His years of back trouble and despondency were physically, ethically, and metaphysically excruciating. They destroyed his youthful ambitions, and with them his self-respect, and left in their wake a shameful personal identification with evil. During the 1870s and 1880s, James constructed for himself a philosophy that was configured, in a multitude of different ways, around these evils of pain and illness and their implications.

Pain and its partner pleasure were intrinsic to his rebuttal of the medical materialists' conviction that we are essentially mechanical beings whose mental activity is wholly determined by the physical body. James deployed hypotheses about the evolutionary role of consciousness and feeling to undermine the assumption that they may be treated as inconsequential or "a mere collateral product of our nervous processes unable to react upon them."[134] This in turn opened up the possibility that his own mental distress could be addressed directly, via psychological rather than physiological interventions.

Metaphysically, his ideas from this period accommodated both the presence of evil but also the existence of a divine plan that would, eventually, see it defeated. He characterized his faith as optimistic, rather than pessimistic, but this optimism was not to be taken for granted and nor did it consecrate the inevitability of evil. In a letter to Renouvier, he explained his conviction that the latter kind of "quietistic fatalistic optimism" led to the antinomianism of his youth that his father had espoused and, in James's words, "some sink of corruption always lies practically at the end." He, on the other hand, proposed a "limited or moral" optimism that was conditional, built on human action rather than passive contemplation and complacency. He held that "we are justified in believing that both falsehood and evil to some degree *need not have been.*" In this context, James submitted that "free will, if accepted at all, must be accepted as a postulate in justification of our moral judgment that certain things already done might have been better done."[135]

For James, this indeterministic faith conveyed a medical imperative. Chapter 2 will explore James's conviction that it was the duty of humankind to strive to conquer the evils of illness and pain and bring about "the highest good," namely, a "proper," healthy life for all. In his eyes, all forms of deterministic or fatalistic philosophy were flawed in that they sanctioned the ethical easy life. Materialism in particular, which James associated with Epicureanism and utilitarianism, hailed the quest for worldly pleasures and the option to numb painful miseries using whatever methods were available, including the ultimate anesthesia of self-destruction. In contrast, a universe of genuine moral relations provided a motive to resist the evil of pain and a stimulus to endure.

CHAPTER 2

Health and Hygiene

The Laws of Health

By the end of 1880, James appeared to have put his most miserable years behind him. He had acquired an assistant professorship in philosophy at Harvard and a contract with Henry Holt & Company for the book that would, eventually, be published as his *Principles of Psychology*. He had also gained a wife, Alice Howe Gibbens (1849–1922), and become a father to Henry, the first of his five children. Ill health was still a problem, however.

Although his back no longer vexed him, James's weak eyes were now the cause of his troubles. Just as he had in his younger days, during the 1880s he escaped to Europe, seeking rest and restoration, and he framed his therapeutic quest as a moral mission. He remained opposed to materialist philosophies but drew heavily on neurological phrases and concepts to explain and manage his suffering, a symptom, it would seem, of his recent acquaintance with the American neurologist George Miller Beard (1839–83).[1] James enthusiastically adopted Beard's novel diagnosis of neurasthenia, a condition of nervous exhaustion brought on by the fatiguing pace of modern American life that manifested itself in an enormously wide range of mental and physical complaints, including all those experienced by James.[2]

James's interest in Beard's work was not confined to his own personal needs, however. He may have long since abandoned all plans to practice as a physician, but medical preoccupations remained a constant presence in his professional life, and neurological ideas were particularly prominent in his writings. In his youth James's thinking twisted itself around the evil of pain. Later, his ideas and research orbited the ideal of health. His veneration of this ideal was complicated, and at times ambiv-

2.1 James's wife, Alice Howe Gibbens James, and their first child, Henry James III, photographed in Boston, circa 1881. (James family additional papers [MS Am 2955 (51),
Box: 1], Houghton Library, Harvard University.)

alent, but the promotion of health became the aim around which many
of James's private and public activities were organized.

In this sense James was very much a man of his time. By the late nineteenth century there was a widespread and growing emphasis on civic
health across America and Europe. The denizens of James's state exemplified this newfound zeal, and the Massachusetts Board of Health was
one of the first such state organizations to be established, in 1869. The

board was given "general oversight of all matters relating to the health of the people" and responsible for making "sanitary investigations and enquiries in respect to the causes of disease, and especially of epidemics and the sources of mortality." It was also instructed to report each year with any recommendations for "legislative acts as they may deem necessary."[3] Alongside this burgeoning state legislation, moreover, existed another set of medical decrees. The "laws of health," as they were known, did not focus on large-scale public health initiatives but on the everyday choices of the individual.

These "laws" comprised the detailed medical guidance that was disseminated via a considerable popular advice literature. These texts assumed and promoted two main hygienic beliefs: first, that the body's natural state is one of health and, second, that an extensive set of "influences" are potentially injurious to the body and its innate healthiness. For James and his contemporaries, the term hygiene pertained to much more than cleanliness and sanitation. Careful attention was paid to the air, food, and drink taken into the body, the regulation of sleep and motion, processes of evacuation, and the passions of the mind. In practical terms the laws of health continued a centuries-long focus on the "non-naturals" of ancient Galenic medicine.[4] In other aspects, however, the mid-nineteenth-century promulgation of these old ideas had its own distinctive character. Medical historians have highlighted the transition to a physiological standpoint and rationale, rather than a simple list of "dos and don'ts."[5] In addition, the advice in question was often freighted with the explicit threat of moral judgment.

The "laws of health" were not legally mandated in the conventional sense, but many felt them to be binding nonetheless. They were laws in two senses: both scientific lore and social imperative. One of the early American tracts, for example, from 1838, was unambiguously titled *Obedience to the Laws of Health, a Moral Duty.*[6] The arch-hygienist, the English philosopher Herbert Spencer, whose work James read at length, was equally bold:

> Few seem conscious that there is such a thing as physical morality. Men's habitual words and acts imply the idea that they are at liberty to treat their bodies as they please. Disorders entailed by disobedience to Nature's dictates, they regard simply as grievances: not as the effects of a conduct more or less flagitious. Though the evil consequences inflicted on their dependents, and on future generations, are often as great as those caused by crime; yet they do not think themselves in any degree criminal. . . . The fact is, that all breaches of the laws of health are *physical sins.*[7]

In America, this sense of ethical obligation evident in the hygienists' program was supported by, and drew strength from, a variety of secular and religious beliefs and movements of the time. Within the communities that James associated with, following the Second Great Awakening, during the early nineteenth century, Calvinist perspectives, preoccupied with predestination and a vengeful deity, were displaced by representations of God as a loving father and the possibility of salvation for all those who worked for it. Although disease and affliction were no longer seen as punitive and inevitable, their avoidance required effort and compliance. The cultural ascendence of natural theology ordained a worldview in which science and religion were mutually supportive. In this context, humankind's God-given powers of reason were to be put to use in the study of the divine wisdom observable in the natural world and the workings of the human body. Physiological research and the daily demands of healthy living could be cast as a Christian duty. To live in accordance with the laws of health was to live in accordance with the wishes of nature's benevolent designer.[8]

In the political realm, the new discipline of medical statistics divided the population into "sustainers" and "dependents" according to their economic contributions to society. Within these kinds of analysis, personal disease and infirmity were positioned as a significant collective problem and a drain on society's resources.[9] In this way, social responsibility and the prevention of sickness were explicitly conjoined. Within the medical profession, moreover, there was common assent to the notion of the hereditability of acquired conditions. An individual who failed to do everything in their power to protect and promote their own health risked their children's fate as well as their own.[10] With the advent of Darwinian ideas these consequences extended further still; the fate of the entire human race was implicated.[11]

James himself was, in his early adulthood, an outspoken proponent of ideas that would come to be associated with eugenics, a term coined by the English polymath Francis Galton (1822–1911) in 1883. Galton had been publishing on related themes since 1865, however, and the theories and values he espoused were in many respects predated by other authors. Concerns about the suitability of marriage partners and prudent reproduction were widespread long before this date.[12] In an 1874 review article on a group of recent medical texts, James made his own position clear, stating that "Darwinism has begun to awaken people's consciences about the responsibilities of breeding in the human race; and we fancy that before long, nervous brides and bridegrooms without jaws or shoulders will be rarer to meet than they unhappily are now."[13] In his comments on one of the new works, *Responsibility in Mental Disease* by Henry Maudsley, he elaborated further on his views, which appear to have been invoked, in particular, by the menace of mental pathology:

> To those who think no evil seems worthy of that name when compared with the evil of insanity we cordially recommend Dr. Maudsley's excellent little volume. In it they will learn that the evil is not wholly beyond human power to avert, and that right breeding is one of the most important means to this effect; although he is less hopeful than we are that the quality of "sexual selection" may be gradually improved by enlightening public opinion on the subject.[14]

James was unconvinced by the need for any regulatory intervention to restrict marriage and reproduction and preferred to trust in the independent choices of an educated society:

> Once expose certain kinds of matches to public reprobation, and the parties will attract each other less. The current ideal of character for each sex is, like other aesthetic ideals, always in part a matter of fashion. . . . And so if reflection, public fashion, the newspapers, and novels all begin to . . . reinstate manliness and bodily soundness on the one hand, and simple loveliness on the other, into the rank they held in the fiction of old, why, it will be strange if some individuals do not fall in love accordingly, and the succeeding generations acquire a tinge more of burliness than would otherwise be the case.[15]

James's gendered notions about the constitution of perfect health in male and female form were in keeping with those of his medical brethren, many of whom went so far as to present women as an appendage to a womb, their health and ideal character subsumed entirely to their function as reproductive agents.[16] In other ways, however, James's reverence for physical health was at odds with many of his contemporaries. His early correspondence indicates that his own reservations about the suitability of particular romantic matches and the hereditability of disease were driven by his concern to eradicate the evils of pain and suffering. In contrast, the medical statisticians mentioned earlier endorsed a very different form of health worship, one based not on metaphysics but on a thoroughgoing materialism.

In another publication from 1874, James discussed this emergent focus on "the political economy of health." James's review of the "Fifth Annual Report of the State Board of Health of Massachusetts" included his comments on a paper by a Dr. Jarvis on this topic. According to James his paper "goes through the usual exhibit of the waste involved in all sickness and in premature death, and accounts a dead child as having been a consumer pure and simple—who has never paid his debt."[17] James ended his own report with a vehement rebuttal of this kind of economic evaluation of life:

> Abstraction made of the moral side, the child is of course nothing but a consumer. But there are forces that no political economist can measure; and while the child lived were it but three years, he was a producer of prudence, industry, and energy in his father. The hours his mother spent caring for him when ill and dying are certainly a pure loss economically, but. . . . what statistician can count the mother's gain of moral depth and force during those very hours, as they raised her from the cheating gates of life up to the gates of eternity? And so of invalidism and old age: they do not only consume, but call forth what but for them would hardly exist, unenvying devotion, grateful piety.[18]

For James then, the existence of a moral facet to life prompted a reappraisal of society's sick and infirm members. In this alternative context, they brought forth a productivity that could be reckoned in a currency that wasn't monetary, and this observation represented a continuation of James's meditations on the significance and purpose of the evils of pain and illness, that were discussed in chapter 1. This alternative ethical perspective, which ran in tandem with his appreciation of physical health as an ideal, was to become a crucial part of James's deliberations, two years later, about whether or not he was fit to enter into marriage. Although his general state of health appeared to have improved somewhat by the time he met Alice, James still viewed himself as an invalid. Their protracted and tortured courtship lasted for two years before James finally committed to marrying her. During this period his correspondence conveyed his misgivings about "the morality of marrying" and their health-related origins:

> There is a natural "plane" of life whose prosperous results justify themselves and therefore refer to no end outside of themselves. It is the open, healthy, powerful, normal life of the world, salubrious, unconscious—the only one that can be in any way *officially* recognized by society and the only one wh. some people think shd. be recognized at all. All departures from the standard of wholesomeness it establishes, are crimes against its law. . . . I for one have an immense piety towards it. Crimes against its law, such as the marriage of unhealthy persons, can only be forgiven by an appeal to some metaphysical world "behind the veil" whose life such events may be supposed to feed.[19]

From James's perspective, only a shared faith in the existence of an un-seen religious dimension had the power to sanction their marriage, which would otherwise be a "*crime* against nature." If such a metaphys-ical world does exist, he continued, "its spiritual ends may run *across* the ends of nature, . . . its laws may defy those of the outward order" or, "in other words," "the deepest meaning of things" may only be revealed with reference to this superadded moral order. Any man who chooses to take this spiritual gamble, James cautioned, "does so at his peril" and should expect "no countenance, perhaps even no tolerance from public opinion."[20]

Although ultimately James pledged his allegiance to the "world 'be-hind the veil'" and married Alice, their correspondence illustrates the depth of his commitment to the sanctity of the natural world and its standards of wholesomeness. In his professional life, James made no se-cret of this "piety" and his devotion to the laws of health. Apart from the implications for marriage and reproduction, he researched and lectured extensively on the hygienic import of the field of physiology. In 1874 he reviewed *Physiology for Practical Use*, edited by the medic James Hin-ton, whose earlier text on pain had impressed him. In his review James acknowledged his own familiarity with the popular medical advice lit-erature and, despite lamenting its lack of scientific rigor, indicated his approval of its underlying ethic. In his final assessment of the Hinton volume James concluded that it was, "as far as it goes, accurate and very readable." He also remarked that "some will object to the frequent mor-alizing of the editor. We ourselves do not."[21] When James first joined the teaching staff at Harvard, in the Department of Physiology, one of the earliest lecture courses he gave was on "Physiology and Hygiene," which ran from 1879 to 1882 and was open to all students.[22] Within this series of lectures there was one subject in particular that appears to have captured both his and his students' attention. In the Harvard archives there is a folder of research articles, collated by James over many years, that is devoted to the controversial topic of alcohol.[23]

The Alcohol Question

Temperance and moderation were, in general, the watchwords of the hy-gienic program, and meat consumption and tobacco use were frequent sources of consternation. Alcohol, however, was the most contested is-sue of all, and there was mounting public feeling at this time that teeto-talism was the only solution.[24] There is only a scanty record of James's hygiene lectures, but his session on alcohol was written up for the *Bos-ton Evening Transcript* in 1881. The article was entitled "Scientific View of Temperance." According to this report, James's lecture dealt with two different kinds of medical evidence marshaled by those arguing for and against moderate drinking, namely, the observations by physiologists of the immediate effects of alcohol upon the body, and "generalizations from experience; e.g. statistics."[25]

Temperance campaigners had, in recent years, seized on claims by French scientists that alcohol was not broken down in the body and asserted that, since these findings proved that alcohol was not a food, it must be a poison.[26] James cited subsequent, contradictory research that, he explained, had shown conclusively that alcohol does get used by the body and can prevent weight loss if less than the usual quantity of food is being consumed, for example, during illness. In ordinary usage he warned, however, that alcohol "rather takes the place of food by reducing the demand for it, this reduction resulting from a diminution in the activity of the vital forces." He concluded that this state of affairs "is not desirable, as it is not consistent with the state of perfect health."[27]

James similarly dismantled the claims that alcohol warms the body or stimulates the muscles. In the first instance he clarified that the use of the spirit merely dilates the blood vessels in the skin, giving the illusory impression of increased heat, when a thermometer will show that the temperature of the interior of the body as fallen. In the latter case he maintained that the seeming increase in activity and vigor is likewise deceptive, and alcohol is "the worst of stimulants for the muscles." He conceded that alcohol was an effective aid to digestion. However, he insisted that since the action here is "strictly that of a flavor to satisfy the nerves of taste . . . the results desired may more safely be secured by the use of other flavors, as, for example, fruits."[28]

James ended his lecture with a survey of the available statistical evidence about the general impact of alcohol on health. The records provided by hospitals and insurance companies, he argued, showed that although whole nations where it had been widely consumed for years were healthier than would be expected if it were a poison, within these societies "teetotallers" were healthier than moderate drinkers. Specifically, he referred to figures indicating that nondrinkers lived longer than drinkers.[29] The reviewer reported the sentiment of James's summing up as follows:

> One may well ask, therefore, in this struggle of modern American life, with its too rapid succession of burdens, is it well to start handicapped by the use of a beverage the effects of which are, on the whole, likely to be injurious? The young man of average physical health will do well to consider whether he can safely adopt such a course.[30]

In his conclusion and his assessment of the physiological evidence, taking into account the balance of hygienic facts, James presented a decidedly negative evaluation of the practice of moderate drinking. There was one notable exception, however, one property of alcohol that James could not dismiss. This was its role as an anesthetic. In this and subsequent lectures on the same topic, James lingered on this aspect, clearly captivated by it.

A series of manuscript lecture notes survive that express in his own words, albeit in abbreviated form, James's feelings about alcohol. He wrote that its "anaesthetic qualities are the secret of its fascination," "anaesthesia the seduction." He also gave more detail about this particular characteristic, explaining that it brought about both "sensorial anaesthesiae" in that "pain—cold—fatigue disappear" and also "cerebral effects."[31] Even in moderate quantities, he clarified: "It obliterates a part of the field of consciousness, and abolishes collateral trains of thought. The association of ideas is less vigorous, and thus are brushed from the mind all the cobwebs of unpleasant memories."[32] Elsewhere he put it more conversationally: "'Drives away care'—'don't care a rap' Worry disappears."[33]

Given James's own prolonged experiences with back pain and melancholy, and a reference in his youthful correspondence to the pleasure he took in drinking "pint after pint of beer," it seems likely that he was speaking from his own experience on this point.[34] This inference is further strengthened by the direction in which James developed these thoughts about the attractions of drunkenness. There are significant limitations to this state, he insisted, as an inebriated individual is "incapable of anything, yet happy." More specifically, he elucidated, "To *work* on alcohol is a most treacherous business."[35] The reviewer of his student lecture reported his views on this point in more detail:

> Experience shows, however, that the work of authors and others that is done under the excitement of alcoholic stimulus does not stand the test of a sober after-judgement. Any task requiring more than half an hour for its accomplishment cannot be so well done under the influence of alcohol as without it. As a rule, then, never take alcohol before or during work; but if at all, when the labor is finished at night, as an anaesthetic to help sound sleep.[36]

James considered it his moral duty to make an earnest attempt to contribute something useful to society via his research. Alcohol, it would seem, may have offered an easily accessible respite from mental and physical pain, but at the cost of him carrying out labor of any worth. In this context, his idiosyncratic introduction of the concept of anesthesia into his early philosophical writings, discussed in chapter 1, takes on a new sense of clarity and mundanity. The only available anesthetics of the time, such as alcohol and opium (which also gets a mention in his lectures), were incompatible with his ethical ideals of civic service and strenuous effort. Out of necessity then, from James's perspective, suffering and endurance were inescapable components of a moral life, whereas anesthesia was allied with what James saw as the value-less and hedonistic nature of materialism.

James's personal reservations about alcohol were part of a much broader fervent of diverse religious, moral, and political agendas.[37] Foremost among these, in the wave of temperance protests that took off during the 1870s, was the domestic plight of women unable to escape the difficult and abusive behaviors of drunken husbands and sons.[38] This was not a theme that James discussed explicitly in public, but during the 1880s and 1890s, as his own brother Bob's drinking became increasingly problematic, James came to appreciate the devastation wreaked in such cases at firsthand.

While in the midst of one his bouts Bob's behavior was extremely erratic, and his wife described his turning up after one such night of drinking in town with no knowledge of how he had cut and bruised his face and "in his most pathetic mood, praying to be confined, saying he was really insane, etc." (In later years, he spent long periods of time living apart from his family as a voluntary resident of various inebriate institutions and asylums.) Bob understood his drinking as an expression of his "morbid irascibility," and, as time wore on, James observed that this tendency appeared to be growing worse, recounting to Alice that "he seems to spend his life mainly between two contrasted emotional states, the irascible before, and the pathetic after, his sprees."[39]

James despaired of the impact of Bob's conduct on his wife and children and, to a lesser extent, on his own Alice, who was the unfortunate recipient of much distressing correspondence from her brother-in-law. Despite being critical of his behaviors, however, James refused to judge him for them. He was resolutely convinced that Bob was "a victim" rather than a villain, and to Bob himself James wrote:

> Now every human being has different temptations, and no one yet acted so ill, but what one acquainted with all the inward sources of his conduct, would have said the action flowed *naturally* from them. Nevertheless, the action as such, as an outward fact, may be bad in the extreme. . . . I think I enter as much as any one into the trials which your nervous system inflicts upon you. Its[*sic*] a daily fight, where

the rest of us have none at all, but it may wear off with age, if you can beat each day.[40]

This notion of "natural" consequences was the crux of the matter for James. He consistently described Bob's intemperance in medical terms and described him as a "hollow shell of a man, covering up mental disease."[41]

It wasn't until the early decades of the nineteenth century that American and European physicians began to conceptualize excessive drinking as a disease rather than a vice.[42] In the latter part of the century the debate these ideas incited was still very much alive and was one to which James made his own contributions. In his review of a text about the infamous criminal Jukes family in 1878, he agreed with the author's opinion that "the temperance question is one for physicians and educators rather than for legislators and politicians," and, in his subsequent public lectures to the Harvard Total Abstinence League in 1886 and 1895, he expanded on this medicalized view of intemperance.[43]

In these addresses James cited the diagnoses of "dipsomania" and "chronic alcoholism," terms that had been adapted into English from the early nineteenth-century works of European physicians. James used alcoholism to denote the habitual use of alcohol that has reached dangerous levels and is accompanied by a "short life" and "degeneration of tissues." Dipsomania was a term that was associated with the translated works of the German-Russian doctor C. von Brühl-Cramer who defined it as an abnormal, involuntary craving for alcohol and argued vociferously that this condition was best treated as an illness. He also maintained, in line with James's brother's interpretation of his own drinking, that the desire for alcohol could arise from strong emotions such as anger and vexation, or sorrow and grief.[44]

James differentiated between two different sorts of alcohol craving: "Congenital & acquired." The first he associated with a "Neurasthenic" diathesis.[45] His lecture notes suggest that he drew heavily on George Miller Beard's portrayal of the desire for alcohol as an understandable product of its apparent action as a stimulant.[46] Those who were unfortunate enough to suffer from chronic nervous exhaustion were, naturally, attracted to the energizing promise of alcohol, along with tobacco, tea, coffee, and "all other stimulants used as whippers up of fatigue." Equally, however, this vulnerable state of bodily depletion need not result from neurasthenia but may also be brought about by an unhealthy environment and lifestyle: "starvation, bad air, insipid food, etc."[47] In both cases, James argued, an uncontrollable need for alcohol could only arise as a result of regular use. In the absence of any experience of, or reliance on, alcohol it could not become a problem. In this way, he concluded, drink craving was a "disease of habit."[48]

Habit

James's writings on habit have attracted a substantial amount of attention over the years, and his chapter on the topic in his *Principles of Psychology* is one of those for which these volumes were best known.[49] Originally published in *Popular Science Monthly* under the title "The Laws of Habit," it became a "popular classic" and was subsequently reprinted as a short book.[50] To understand why the concept of habit was so important to James, however, it is essential to locate these texts within the contemporary hygienic program to which he was deeply committed.

From James's perspective, habits and health were linked in a variety of different ways. In their simplest association, bad habits, namely, personal behaviors that contravened the laws of health, led to disease and untimely death. Excessive alcohol use was high on this list of dubious proclivities, along with masturbation, which was regularly linked with the development of insanity in the psychiatric literature.[51] In a letter to one of his sons, James warned him of this danger:

> Now that you are . . . exposed to all sorts of bad boys older than yourself, I ought to give you a word of moral advice. You'll hear a lot of dirt and smut talked, and if you have a taste that way, you can talk it yourself as much as you please, and no one will prevent you. . . . If any boys try to make you *do* anything dirty, though, either to your own person or to their persons, it is another matter, and you must both preach and smite them. For that leads to an awful habit, and a terrible disease when one is older. So don't stand that for a single instant in your presence, and you may save some other boy from a bad future.[52]

Within the hygienic literature moral reprobation was not reserved for such notorious vices and extended to a far greater number of daily habits. The minutiae of everyday life were subject to the laws of health, and those who were dedicated to this "physical morality" were advised to consult the legislation before taking a bath, opening a window, or having a snowball fight, lest they inadvertently sin.[53] Within this medical paradigm the smallest departure from wholesome routine, even something as seemingly innocuous as an unattended accumulation of earwax, could lead to physical debility.[54] James, on one occasion, berated himself for partaking of a single cup of coffee.[55] Beyond these hygienic dictates, however, habit was more profoundly implicated in James's understanding of health and disease.

In the opening passages of his famous chapter on habit, James determined the founding principle of the doctrine that followed. This "first proposition" was, namely, that *"the phenomena of habit in living things are due to the plasticity of the organic materials of which their bodies are composed."* James conceived of habits as acts that had become physically embedded in an organism's tissues. He then went on to explain that, owing to this organic plasticity and the "inertia of the nervous organs, when once launched on a false career," many types of disease are revealed to be little more than bodily habits that, if they can be arrested once, need never return:[56]

> We find how many so-called functional diseases seem to keep themselves going simply because they happen to have once begun; and how the forcible cutting short by medicine of a few attacks is often sufficient to enable the physiological forces to get possession of the field again, and to bring the organs back to functions of health. Epilepsies, neuralgias, convulsive affections of various sorts, insomnias, are so many cases in point.[57]

According to James there were significant classes of disease that were not caused by irremediable organic defects but were instead a product of physiological habit and, therefore, potentially curable. Also notable is the implication, characteristic of the hygienic discourse, that left to its own devices in the absence of bad habits the body will heal itself, or, in other words, the body's natural state is one of health.

In the remainder of the chapter James went on to explore the mechanism and significance of habit formation in more detail. Here he relied heavily on one particular text: William Carpenter's *Principles of Mental Physiology, with their Applications to the Training and Discipline of the Mind and the Study of its Morbid Conditions,* a book that James reviewed three times on its publication in 1874. William Carpenter (1813–85) was an English physician who, instead of practicing medicine, became a physiologist and published several influential texts during the middle decades of the nineteenth century. He intended his *Principles of Mental Physiology* to extend the field of "Psychological Science" into the neglected "border-ground" between physiology and metaphysics, inquiring into both "the action of Body upon Mind, as well as of Mind upon Body."[58]

One of the central themes of Carpenter's book was, reported James, "to trace how acts and thoughts originally voluntary become, by the force of habit, automatic, by the aptitude of the organization to *grow* to the modes in which it has been much exercised, and probably to transmit to its offspring the same tendency." In short, Carpenter depicted a psycho-physiological system where volitional effort, or a "persistent will," eventually physically "moulds the organism to its purpose."[59] In the chapter of Carpenter's book entitled "Of Habit," the author explained that the human body, and in particular the "ganglionic substance of the Brain," is capable of constant regeneration, and this reproduction of new nerve tissue facilitates the modification of the organism. Carpenter pointed to habits, such as walking erect, which are acquired and become automatic, ingrained as a "Nervous Mechanism."[60]

Carpenter contended, moreover, that what was true for skills of dexterity, such as walking, was also true for thought processes: "every state of ideational consciousness which is either very strong or is habitually repeated, leaves an organic impression on the Cerebrum; in virtue of which that same state may be reproduced at any future time, in respondence to a suggestion fitted to excite it." Fundamentally, Carpenter argued that recurring patterns of intellectual and moral activity (in the nineteenth-century sense of the word "moral," meaning pertaining to the will or emotions) become established as habits and, ultimately, dependent on a "reflex action of the Cerebrum." In this way our characteristic psychological responses become written into the physiology of our nervous system.[61]

James's writings on habit imply that it was this notion, that our characters and not just our physical skills develop out of our repeated efforts over time, that constituted the primary appeal of Carpenter's ideas. For James, the real potential of the physiology of habit formation lay in the prevention of one particular category of disease, that of insanity. When he first reviewed Carpenter's *Mental Physiology* text, he did so in connection with the topic of "mental hygiene."

James's review article "Recent Works on Mental Hygiene" was revealing in that it evidenced his wide reading and considerable interest in this topic.[62] One of the key texts in the oeuvre, which he mentioned, was written by Isaac Ray (1807–81), who was, in James's opinion, a writer "easily at the head of" the "honorable profession" of alienism in their country.[63] Mental hygiene comprised, according to Ray, "the art of preserving the health of the mind against all the incidents and influences calculated to deteriorate its qualities, impair its energies, or derange its movements." In many ways his subsequent disquisition followed the conventional narrative of the general hygiene manuals, and the usual admonitions with regard to exercise, rest, food, clothing, and climate were all present. In addition, the book particularly stressed the significance of "the government of the passions" and "the discipline of the intellect." Both of these, he argued, play a vital role in mental health.[64]

Although the term "mental hygiene" only appeared for the first time in the American medical literature in 1843, many of the principles invoked by Ray in his text form part of a much older tradition.[65] This historical debt was acknowledged explicitly in a book by the Austrian physician Ernst von Feuchtersleben (1806–49), entitled *Zur Diätetik der Seele*, which was first translated into English from the German original in 1852 as *Dietetics of the Soul*. James and Ray both regarded this text as an important contribution to the mental hygiene literature.[66] Indeed, this was one of two more established texts, including Ray's, which James cited in the opening paragraph of his article. While, however, Feuchtersleben's book contained didactic maxims that were consistent with Ray et al.'s, it exhibited a very different overall character.

Feuchtersleben had trained as a doctor, but his book did not resemble the typical hygienic tomes with their reverence for physiological knowledge. Instead, it paid its respects to a different tradition of learning and, according to the editor of an early American edition, positioned itself as a work of "practical moral philosophy." The introductory chapter set out the aims of the text and comprised an homage to the art of self-control. According to Feuchtersleben no art was so neglected by man despite it being the "first and the last of his moral duties" and indispensable to the health of mind and body.[67] He advocated, as one of his main tenets, the management but not the indiscriminate suppression of the passions, and he held up the Stoic *Meditations* of Marcus Aurelius as a source of inspiration for those seeking to follow his dietetic regime.

Feuchtersleben's reverence for the Stoic goal of self-mastery as the keynote of good mental health was shared by several contemporary medical authors. In James's review of the English alienist Henry Maudsley's *Responsibility in Mental Disease*, he pointed out that Maudsley's words, on an individual's ability to resist the onset of some types of insanity via their powers of "self-formation," closely resembled the moral ideas of a "great Roman Emperor."[68] This would appear to be a reference to Maudsley's insistence that the highest accomplishment of "self-development" was the formation of a character where the "thoughts, feelings, and actions are under the habitual guidance of a well fashioned will." The perils of ignoring this task of self-development were severe, he warned, since, left un-checked, the "injurious operation of . . . painful emotions . . . often make shipwreck of the mental health."[69] James noted the alienist's emphasis on the need to develop inward consistency of thought and action, and indifference to outward fortune, and concluded that "moralists need not be anxious when the most advanced positivism comes to practical conclusions that differ so little from those of the 'metaphysically' minded Marcus Aurelius."[70]

It is evident that James had a great deal of personal sympathy with Maudsley and Feuchtersleben's general approach to life and health, and his own affinity for the Stoic philosophy of Aurelius was acknowledged earlier.[71] While James applauded Maudsley's conclusions, however, he was critical of the path by which he had reached them. He accused the author of failing to erect a "rigid and adequate logical bridge" between his philosophical allegiances and his didactic maxims. According to James, Maudsley did not succeed in reconciling his twin commitments to the evolutionism of Spencer and the ancient traditions of self-discipline espoused by moral philosophy.[72] In this respect, however, Carpenter's system received a warmer reception from James. Carpenter's bridge between his own physiological and ethical axioms was, James explained, "the great law of habit." James's cited his words on this point at length:

> We have thus a definite Physiological *rationale* for that "government of the thoughts," which every Moralist and Religionist teaches to be the basis of the formation of right character . . . And the Writer cannot but believe that there are many upon whom the essentiality of Intellectual and Moral *discipline* will be likely to impress itself with greater force, when they are enabled thus to trace out its Physical action, and to see that in the Mental as in the Bodily organism, *the present is the resultant of the past*; so that whatever we learn, think, or do in our Youth, will come again in later life, either as a Nemesis or as an Angel's visit.[73]

In the context of James's review article there can be no doubt that the "Nemesis" in question assumed the form of insanity. Over the course of his chapter "Of Insanity," toward the end of his text, Carpenter made it clear that he too considered the possession of an adequate "habit of Self-control" to be the hallmark of a "well-balanced Mind." The "raving madman," on the other hand, is someone whose feeble will is powerless, leaving them the "sport of uncontrollable passion." According to Carpenter that which "is common to every form of Insanity, which is frequently its first manifestation . . . is *deficiency of volitional control* over the current of thought and feeling, and consequently a want of self-direction and self-restraining power over the conduct."[74]

Regarding the causes of insanity, Carpenter acknowledged that heredity, "mechanical injuries of the Brain" and the presence of disease were all significant factors. He also proposed that, in other instances, individuals are, to a certain extent, culpable for their own lapses from sound mental health. He considered cases such as those "generally termed Moral Insanity," which are "particularly common among females of naturally 'quick temper,' who, by not placing an habitual restraint upon themselves, gradually cease to retain any command over it." Principally, the operative cause of insanity in these cases was, Carpenter believed, "the habitual indulgence of an originally bad temper." He also discussed "impulsive insanity" when an individual may be driven, by irresistible impulses, to kill, to commit rape, or to steal and all without any malicious feeling toward the persons injured. Once again he sounded a note of personal responsibility and insisted that such states of mind are "particularly liable to be induced in persons who habitually exercise but little Volitional control over the direction of their thoughts."[75]

Finally, at the close of his chapter, Carpenter explicitly tackled the question of the prevention and cure of insanity. In many cases, he insisted, timely efforts at self-control may succeed in keeping down "exaggerated emotions," which often lead toward erroneous interpretations of events and, ultimately, insane delusions, and succeed instead in directing the thoughts into another, healthier channel. He maintained that there can be "no doubt that many a man has been saved from an attack of Insanity, by the resolute determination of his Will not to yield to his morbid tendencies." He also asserted that, when the case is further advanced, "moral treatment," which consists of the judicious encouragement and nurturing of "volitional direction" by a physician, is often efficacious in bringing about a reacquirement of self-control and thus, the recovery of the patient.[76]

Carpenter conceded that detractors may point out that his ideas are contradictory. If, as he contends, insanity is the expression of a "disordered physical action of the Cerebrum," it might appear inconsistent to expect a man to control this by any effort of his own or to expect any good results from an exclusively "moral" treatment. He maintained, however, that the crucial, conciliating factor is the power of the individual to control which habits of mental action become established, and thus embodied, as automatic physical reactions:

> Whilst the disordered physical action of the Cerebrum, *when once established,* puts the automatic action of his mind altogether beyond the control of the Ego, there is frequently a stage in which he has the power of so directing and controlling that action, as *to prevent the establishment of the disorder,* just as, in the state of perfect health, he has the power of forming habits of Mental Action, [conducive to the ongoing health of] the Brain.[77]

In this way Carpenter used his theory of habit formation to construct a melioristic narrative with regard to the "Nemesis" of insanity.
It was for this reason that James included his *Mental Physiology* volume
in his survey of recent works on mental hygiene. Despite the fact that
the author never once used this term, James insisted that it was a "more
important manual of mental hygiene" than others which bear this title.
Carpenter's stress on the malleability of the nervous system provided a
fertile paradigm within which the brain could be molded and fortified
against the onset of insanity, all with intellectual impunity. In short, the
concept of habit offered James a means by which he could subvert the
fatalism of the medical materialists. In Carpenter's world, a person's nervous machinery, their all-important powers of self-discipline and therefore their destiny, could be bent to their will via the iterative processes
of habit formation. In addition, James pointed out, referring to the La-
marckian theories of hereditability that were authoritative at this time,
this molding process potentially had beneficial consequences for the
health of the next generation too, "in so far as the tendency to habits
thus organized may be inherited."[78]

Integral to Carpenter's theory of habit formation was the emphasis he
placed on youthful experiences and learning. He stressed that it was the
job of parents and teachers to facilitate the "healthful development" of
mental habits in their charges during the window of early physiological
plasticity. He cautioned that whereas the nervous system is particularly
impressionable during its development: "From the time that the Brain
has attained its full maturity, the acquirement of new modes of action,
and the discontinuation of those which have become habitual, are alike
difficult."[79] James took Carpenter's warning to heart, and in each of his
three separate reviews of Carpenter's *Principles of Mental Physiology*, it
was "clergymen and educators" to whom he primarily recommended
the text. In summing up one of these reviews he decreed the book to be
"one of the most valuable pedagogic publications of modern times," and
when James came to make his own contribution to this field, he made
Carpenter's ideas central.[80]

Talks to Teachers

James originally composed his *Talks to Teachers* at the behest of his em-
ployer, Harvard University, as a result of the university's increasing interest in pedagogical matters. Along with their appointment, in 1891, of
an assistant professor of the history and art of teaching, the university
asked instructors, in other departments, to offer relevant courses. As

a result, James first delivered a series of psychological lectures on this topic to an audience of Cambridge teachers in 1892. He subsequently took these lectures "on the road" to derive an income from them, and later on they were published, in installments, in the popular periodical *Atlantic Monthly* and eventually printed together in book form.[81]

Although the lectures sold extremely well, James himself was not overly enamored with either his series of *Talks to Teachers* or the teachers he was talking to. In private he derided their general lack of intellectual agility: If a teacher "does ever understand anything you say, he lies down on it with his whole weight like a cow on a doorstep so that you can neither get out or in with him."[82] In public he voiced his trepidation that he had nothing very novel to tell them anyway, and his first lecture opened with a disclaimer. The editors of education journals and "the arrangers of conventions" had been doing their best, he contended, to conjure up a portentous, and mystifying, aura around "the new psychology" and its implications for the profession of teaching. James confessed that, in his humble opinion, there wasn't much that was genuinely new within the field of psychology, and, moreover, there was a limit to how helpful it could be anyway since "psychology is a science, and teaching is an art." Accordingly, the most psychology can do is lay down "lines within which the rules of the art must fall." In addition, he continued, the "alpha and omega of the teacher's art," those skills of "tact and ingenuity," are "things to which psychology cannot help us in the least."[83]

Regarding the actual material in the lectures, James unashamedly recycled sections of his own, recently published *Principles of Psychology*. A letter to fellow American philosopher and educator William Torrey Harris implies that he had only embarked upon the project reluctantly, at the urging of his university employers. James wrote that "they are forcing me to give ten lectures ... on 'Topics of Psychology of interest to Teachers.' It is lamentable work!"[84] After touring various summer schools with the lectures over the following years, he became even less enchanted with the material, confessing to another correspondent that it would soon become "intolerably tedious."[85] To take James's own assessment of his *Talks* at face value would be a mistake, however. They are interesting in their own right, not because they contain any new examples of Jamesian psychologizing but for what they reveal about his values and ideals at this point in his life. In these lectures, James's hygienic concerns are a consistent presence.

In part James's focus reflected a widespread contemporary concern with "school hygiene." Many nineteenth-century doctors and alienists were troubled by the potential role of education in bringing about insanity and nervous disease. In the edition of Daniel Hack Tuke's *Dictionary of Medicine* that was published in 1892, the same year that James first delivered his *Talks to Teachers*, the entry for "Prevention of Insanity" discussed education in some detail. The author, English alienist George Fielding Blandford, raised the specter of the harmful emotional strain occasioned by scholastic competition. He described how the fierce rivalry for scholarship places, in the case of those with a nervous inheritance, often brought about the "evil consequences of mental disappointment and a sense of failure after years of brain work with all its dangers."[86]

The American neurologist Silas Weir Mitchell also wrote at length about the impact of schooling in his text on the perils of different forms of nervous strain.[87] Meanwhile, Isaac Ray, the American alienist whose work James reviewed, was critical of any pressure being put on children to learn at a very young age and actively discouraged the fostering of academic precocity:

> It is the law of the animal economy that the various organs do not arrive at their full maturity of vigor and power, until some time after the adult age has fairly commenced. To suppose the youthful brain to be capable of an amount of task-work which is considered an ample allowance to an adult brain, is simply absurd, and the attempt to carry this folly into effect must necessarily be dangerous to the health and efficiency of this organ.[88]

James, however, was not interested in merely avoiding undue nervous strain and mental pressure. His hygienic aims were far more ambitious. James envisaged the pedagogical project as an opportunity to actively develop and nurture citizens who conformed to the ideal of health.

In his lectures, James, in his own words, adopted a "biological conception" of human life and education.[89] While he reassured his audience that "no one believes more strongly than I do that what our senses know of as 'this world' is only one portion of our mind's total environment," he also insisted that this immediate world of the senses is "the primal portion" and the "*sine qua non* of all the rest." Without the human mind's ability to produce the "more practically useful results" that are required to survive in this immediate sensual world, he argued, it would not be in a position to sustain those "ethical utopias, aesthetic visions . . . and insights into eternal truth" that pertain to matters beyond it.[90] James privileged, as he saw it out of practical necessity, education as the means by which human beings can optimize their biological resilience. James described the aim of education as in the last analysis . . . the organizing of *resources* in the human being, of powers of conduct which shall fit him to his social and physical world. An "uneducated" person is one who is nonplussed by all but the most habitual situations. On the contrary, one who is educated is able practically to extricate himself . . . from circumstances in which he never was placed before.[91]

He spelled out what this meant for the teacher, who was to regard their professional task

chiefly and essentially in *training the pupil to behaviour*; taking behaviour, not in the narrow sense of his manners, but in the very widest possible sense, as including every possible sort of fit reaction on the circumstances into which he may find himself brought by the vicissitudes of life.[92]

James made it clear that, when he defined education in terms of learning how to extricate oneself from an awkward situation, he was not talking about how to turn down an unwanted invitation to lunch. In the archival notes that relate to his *Talks to Teachers*, he referred to an educated person as one who is "adapted to *emergencies.*" A few lines later he reiterated this point, stating that "adaption to the environment is the watch word."[93]

James's pedagogical standpoint owed much to the evolutionary philosophy of Herbert Spencer, and his lecture notes acknowledged this intellectual debt. There he stated that, with reference to the aims of education, "Spencer has renovated psychol[ogy]."[94] In his essays on *Education* (1861), Spencer specified unequivocally that the primary educational objective should be "direct self-preservation," which included guarding the body against mechanical damage, disease, and death. All other educational aspirations depended upon this one, he added, as "without health and energy, the industrial, the parental, the social, and all other activities become more or less impossible."[95] Spencer also maintained that "self-preservation" was synonymous with his notion of "correspondence," namely, the ability of an organism to adapt itself continually to its changing environment. According to his formulation, perfect correspondence was equivalent to perfect health, or the "highest conceivable degree of vital activity."[96] For him, and for James at this point in time, education, adaptation to the environment, and health were fused together in a hygienic triumvirate.

Despite their commonalities, however, the practical details of James's and Spencer's pedagogical visions differed. Spencer objected to the long-standing belief that a classical education afforded the only suitable preparation for "the learned professions, public life, and cultivated society."[97] He insisted instead that the sciences should take pride of place in the curriculum. Above all, he insisted, scholastic study should concentrate on the science of physiology. A thorough knowledge of the "laws of life," rather than Latin and Greek, was essential, he argued, if humanity was to avoid the perils of disease and untimely death:

> When a mother is mourning over a first-born that has sunk under the sequel of scarlet-fever—when perhaps a candid medical man has confirmed her suspicion that her child would have recovered had not its system been enfeebled by over-study—when she is prostrate under the pangs of combined grief and remorse; it is but a small consolation that she can read Dante in the original.[98]

James, in contrast, did not enter into this debate about subjects and curricula. His focus was not on teaching children the facts of physiology, but rather on helping teachers apply some of these facts themselves, to develop and promote the neurological health of the students in their charge. He maintained that "the great thing in all education is to *make our nervous system our ally instead of our enemy.*"[99] In a passage that was strikingly reminiscent of his article on the mental hygiene insights of Carpenter and Maudsley et al. from many years earlier, James warned his audience about the implications of "the physiological study of mental conditions" for society's children:

> Could the young but realize how soon they will become mere walking bundles of habits, they would give more heed to their conduct whilst in the plastic state. We are spinning our own fates, good or evil, and never to be undone. Every smallest stroke of virtue or of vice leaves its never-so-little scar. . . . Down among [the] nerve-cells and fibres the molecules are counting it, registering and storing it up. . . . Nothing we ever do is, in strict scientific literalness, wiped out.[100]

In terms of which specific habits were to be encouraged, James advised that "the exercise of voluntary attention in the schoolroom must . . . be counted one of the most important points of training that take place there."[101] At first sight this maxim appears to be an innocuous commonplace, referring to the scholarly advantages conferred by superior powers of concentration. A closer reading, though, suggests that it related to the emphasis on the inculcation of self-control that was a consistent element across the contemporary mental hygiene literature. More specifically, James appears to have drawn heavily from the chapter on "The Prevention of Insanity" in Maudsley's *Responsibility in Mental Disease*. James had reviewed the text twice, and on one occasion declared this chapter to be "original and valuable" with respect to its contribution to "the hygiene of the mind" and bemoaned the fact that it couldn't be "reprinted as a tract, and dispersed gratis over the land."[102]

Crucial to Maudsley's recipe for the ideal education, which he laid out in his book, was the cultivation of "an internal power of withdrawing the attention" from destructive emotions such as "anxieties and apprehensions, disappointed ambition, envies and jealousies" and of "fixing it on other and more healthy trains of thought." He used a monetary metaphor to explain why these "depressing passions" are exceedingly dangerous to mental stability. Together with the thoughts "which they call up and keep active in the mind," they involve a "large expenditure of nerve force," he cautioned, comparing this nerve force to savings in an account. Left unchecked, "there can be in the end but one result—insolvency. Slight excesses of expenditure over income, in vital as in financial matters, . . . are cumulative, and inevitably tell their tale at last." Developing the power of self-government may be a "hard and long and weary" mission, he conceded, but it will "vastly lesson the sum of insanity upon earth."[103]

This advice, and the financial nature of the metaphor used to illustrate it, is markedly akin to James's own guidance, which he included in his chapters on "Habit" in both *The Principles of Psychology* and *Talks to Teachers*. In the latter text he stated that his "fifth and final practical maxim" is to keep "the faculty of effort alive in you by a little gratuitous exercise every day." He insisted on the importance of practicing, even when it seems unnecessary, the habits of "concentrated attention, energetic volition and self-denial in unnecessary things." Such "asceticism" is, he explained, "like the insurance which a man pays on his house and goods. The tax does him no good at the time, and possibly may never bring him a return. But, if the fire *does* come, his having paid it will be his salvation from ruin." In "the hour of dire need" these habits, which strengthen our "faculty of effort," are all that stand between a man and his "softer fellow-mortals" who will be "winnowed like chaff in the blast."[104]

The requisite development of these powers of "concentrated attention, energetic volition and self-denial in unnecessary things" was, however, a delicate task. The impression left by James's writings is that these activities all represented a flexing of "the will." There is also a sense, that is never made explicit, that the will is like a muscle that must be regularly exercised to keep it strong, but not strained by unnecessarily arduous use. In James's lecture on "Attention" he discussed the potential hazards associated with less sophisticated methods of keeping the children engaged with the lesson in hand. Rather than "commanding it in loud imperious tones" the teacher should, wherever possible, endeavor to keep their spontaneous attention by bringing out new and varied aspects of the subject matter. The alternative of resorting to "too frequent appeals to voluntary attention of the coerced sort . . . is a wasteful method bringing bad temper, and nervous wear and tear."[105]

Later on, in his lecture on "The Will," James explained the neurological consequences of such uninspired teaching styles. There he returned to this situation, of a class of pupils whose attention is wandering, and recast his previous advice in the medical language of "inhibition." In the chapter on "Will" in his *Principles of Psychology*, he quoted, at length, a passage from the Scottish alienist T. S. Clouston's *Clinical Lectures on Mental Diseases*. Clouston (1840–1915) stated that the "physiological word inhibition can be used synonymously with the psychological and ethical expression self-control, or with the will when exercised in certain directions."[106] In his *Talks to Teachers*, James explained how all "ideas," be these intellectual or emotional states of mind, have the power to produce or check certain activities. In this sense, any notion may be labeled either impulsive or "inhibitory," and it is possible to conceive of all voluntary action as "*a resultant of the compounding of our impulsions with our inhibitions.*" In the case of the teacher who wishes to prevent the minds of their pupils from drifting off into other lines of thought, he stated that there are two different kinds of inhibition that they may utilize:

> We may call them inhibition by repression or by negation, and inhibition by substitution respectively. . . . In the case of inhibition by repression, both the . . . impulsive idea and the idea that negates it, remain along with each other in consciousness, producing a certain inward strain or tension there: whereas in inhibition by substitution, the inhibiting idea supersedes altogether the idea which it inhibits, and the latter quickly vanishes from the field.[107]

James clarified that the teacher who resorts to forcefully demanding their pupil's attention relies on "inhibition by repression," and the resulting state of "inward strain or tension" unduly taxes the pupils' precious stores of nervous energy.[108] Similarly, he insisted, if an individual pupil is balking at a particular problem or task it is not a good idea to compel them to complete it. Rather than a punishable act of moral culpability, such a case should be treated in "nineteen times out of twenty" as one of "neural pathology" whereby "an inhibiting sense of impossibility remains in the child's mind" and they cannot "get beyond the obstacle." The solution in such situations is, he insisted, to leave the problem alone and return to it later when the inhibitory notion has lapsed from the pupil's mind. The more traditional alternative of breaking the child's will is, James warned, "always a scene with a great deal of nervous wear and tear on both sides."[109]

James returned to this concept of inhibition and its potentially pathological effects several times over the course of his career, including, most notably, in his lectures on *The Varieties of Religious Experience*, which are discussed at length in chapter 4. Inhibition was also a feature of a standalone lecture that he addressed to students. In most editions of his *Talks to Teachers* James's pedagogical lectures are followed by three "Talks to Students on Life's Ideals." In the preface to this collected volume, he stated that only one of these three additional addresses, entitled "The Gospel of Relaxation," "properly continues the series of *Talks to Teachers*."[110] In his opening paragraph to this particular address, James explained that his purpose was to

> take certain psychological doctrines and show their practical applications to mental hygiene, — to the hygiene of our American life more particularly. Our people ... are turning towards psychology nowadays with great expectations; and, if psychology is to justify them, it must be by showing fruits in the paedagogic and therapeutic lines.[111]

James's comments confirm that the intended theme of both his *Talks to Teachers* and "The Gospel of Relaxation" was mental hygiene. Although the latter 1899 address has received relatively little attention from contemporary scholars, it provides a revealing context within which to place his well-known and controversial ideas about the production of emotional states and contains important insights into James's developing ideas about mind-body relations. It also includes passages that help to refine what "health" meant to James at this point in time.

Emotions and the Body

Central to the first section of "The Gospel of Relaxation" is James's much-debated theory of the emotions. This theory emphasized the bodily dimension of our emotional responses and contended that these responses are "mainly due to those organic stirrings that are aroused in us in a reflex way by the stimulus of ... [a] situation" rather than a direct effect of this situation on the mind.[112] In his original 1884 essay "What Is an Emotion?," he expounded his thesis in more detail. There he wrote:

Common sense says, we lose our fortune, are sorry and weep; we meet a bear, are frightened and run. . . . The hypothesis here to be defended says that this order of sequences is incorrect, that the one mental state is not immediately induced by the other, that the bodily manifestations must first be interposed between, and that the more rational statement is that we feel sorry because we cry, . . . afraid because we tremble.[113]

Superficially, at least, this observation seems out of place among James's other writings. It appears to imply that mental activity is wholly determined by the state of the body, a position that elsewhere James had vociferously opposed. A private letter, however, which he sent in 1884 shortly after his original essay on the emotions was published, discloses that James never intended his theory to be indicative of a materialist philosophy of mind:

Of course my "theory" about the emotions is quite independent of what species of mental operation the emotion is consequent upon,— whether a reflective judgement or a "perception." It confines itself to saying that, when once there, the emotion consists of a complex of feelings of bodily change. I used the word perception untechnically and for brevity's sake to cover all sorts of cognition.[114]

James's subsequent "Gospel of Relaxation" lecture implied that his motive for stressing the physicality of emotional experience was practical, rather than philosophical. More specifically, it was a product of his health-oriented mind-set. In this text his theory of emotions was presented explicitly as a hygienic tool; emotional understanding was the first step toward the medically important task of emotional management. He maintained that "the mere giving way to tears, for example, . . . will result for the moment in making the inner grief . . . more acutely felt." Consequently, he insisted, there is, a pedagogic maxim to be drawn from this observation:

There is . . . no better known or more generally useful precept in the moral training of youth, . . . than that which bids us pay primary attention to what we do and express, and not to care too much for what we feel. . . . Action seems to follow feeling, but really action and feeling go together; and by regulating the action, which is under the more direct control of the will, we can indirectly regulate the feeling, which is not.[115]

James then proceeded to expand this connection between bodily actions and states of mind into a more general "law of psychology," namely, that every physical act, small or large, contributes to a "ceaseless inpouring [of] currents of sensation . . . which help to determine from moment to moment what our inner states shall be." Over the course of his lecture, he developed the significance of this law with respect to everyday behaviors and the maintenance of mental health. He referenced the observations of foreign visitors to America who had commented on the physical tension present in the faces and bodies of its countrymen and countrywomen, characterizing this as a "wild-eyed look . . . either of too desperate eagerness and anxiety or of too intense responsiveness and good-will." "American over-tension and jerkiness and breathlessness and intensity and agony of expression" may, he lamented, currently be admired as displaying vivacity and intelligence, but instead they should be seen for what they really are: "*bad habits*" that result, he argued, in frequent and severe nervous breakdowns.[116]

He explained that the "sensations that so incessantly pour in from the over-tense excited body" result in an overtense mind and an ever-present "threatening" and exhausting "inner atmosphere." The remedy, James continued, lies in the "origins of the disease." To achieve inner calm, we must practice outer calm, deliberately relaxing our muscles and moderating our breathing. Here he introduced the work of Annie Payson Call (1853–1940), a woman whom he described to his wife as "awful," meaning awe-inspiring rather than dreadful as this adjective would be understood today. James insisted that her *Power through Repose*, which promoted the healthful benefits of physical relaxation, is "a book that ought to be in the hands of every teacher and student in America of either sex."[117] In his earlier review of the book he had recommended that a "philanthropist with some surplus income might do a much worse thing than to distribute it widely among the teachers of the land."[118]

James developed his argument further still, insisting that, in terms of mental hygiene, the ideal body is not only relaxed, but also "well-toned." He mentioned the revolutionary impact of the new fashion for skiing in Norway that had seemingly transformed the delicate constitutions of the Norwegian women and with them their mental characters. He hoped that the contemporary American enthusiasm for tennis, tramping, and bicycling would likewise lead to a "sounder and heartier moral tone" among the female members of his society.[119] James practiced what he preached and considered physical activities, such as hiking, to be an essential part of life for him and his children. In a letter to his daughter, Peggy, he wrote: "It is jolly that you like your games now. That is the basis of soul as well as body. . . . In a few years you will probably be invited to the Adirondacks, and get initiated into the hill climbing there, to which I truly believe that I owe more than to any other influence of my life."[120]

In one respect this theme was not a novel one within James's writings. His assumption that a strong, healthy body fosters a healthy mind was also present in his much earlier discussion of the mental hygiene literature. In this sense he grafted his later emotional theorizing onto a belief that long predated it in his thinking. In his 1874 review article, for example, James took umbrage with D. A. Gorton, the author of *An Essay on the Principles of Mental Hygiene*. "The wisdom of some of the author's ideas is," he argued, "more than questionable." He referred his readers to the section of the text where Gorton claimed that steam, electricity, and mechanical inventions had effectively superseded the need for physical strength and development. Gorton argued that when contemporary society relied solely on the brain and nervous system "backed by creations of mechanical art and skill," to perform tasks that were previously undertaken by manual labor, it was questionable whether humankind should continue to maintain the degree of muscularity that was historically necessary for earlier civilizations.[121] James disagreed:

> We believe, on the contrary, that the indirect benefits we shall get from "keeping up our muscle" . . . will be quite as important as its immediate mechanical uses have ever been; and that, if our race ever gets to disregard muscular strength as an element of human perfection, it will be far gone on the path of degeneration. A well-developed muscular system means something more than the ability to raise a given weight. It is apt to be correlated with calm nerves and strong digestion, with joyousness and courage of every sort, with amiability and all manner of sound affections and sensibilities.[122]

James went on to acknowledge the "immense debt" that society owes to its members with "nervous temperaments, its Shelleys, its J. S. Mills, and others." He reiterated, however, that, in terms of a general type, the "old vascular and muscular breadth is better for the stock of human nature, and we trust that here in America we shall never lose the wholesome sense of this."[123]

Twenty-five years later and the opinions James voiced in his "Gospel of Relaxation" were much the same. He even mentioned Gorton's thesis, (although he claimed to have forgotten the name of the author), and continued to insist that "muscular vigor" will never be a superfluity.

Quite apart from its "mechanical utility," it is, he claimed, "an element of spiritual hygiene of supreme significance" and will always be needed to "furnish the background of sanity, serenity, and cheerfulness." To this end, he hoped that, in America, the "ideal of the well-trained and vigorous body" will be maintained "neck by neck with that of the well-trained and vigorous mind as the two coequal halves of . . . higher education."[124] (Here again James described the driving aims of education in terms of health using, in this specific instance, the synonym "vigor.")

James was far from alone in espousing such ideals, and the healthy body, healthy mind refrain was commonplace within his late nineteenth-century world.[125] Ray, for example, in his text *Mental Hygiene*, was adamant that ill health is "one among the most efficient causes of insanity." He proposed that there are two requisites for a sound and vigorous mind: a brain free from "disease and deterioration" and the "healthy condition of the other bodily organs."[126] From a philosophical perspective, however, James's commitment to this sentiment is intriguing, as it offers an insight into his developing thinking about the mind-body problem. At this point in time, James upheld this faith, that physical health is a prerequisite for optimum mental health, while also, increasingly, looking to theories about the psychological causation of mental and physical disease. In this sense he maintained two symmetrical investigatory medical paradigms, and it appears that it may have been this commitment to a dualistic etiology that endeared him to the work of the Austrian neurologist Moriz Benedikt (1835–1920).

In "The Gospel of Relaxation" James introduced Benedikt's concept of the "Binnenleben," which he translated as the "buried life of human beings."[127] James explained that, according to Benedikt, it was obligatory for any doctor hoping to "get into really profitable relations with a nervous patient" to grapple with this inner realm. James recounted how Benedikt's concept of a "buried life" or "unuttered inner atmosphere" describes the "inner personal tone" that we can't communicate to others "but the wraith and ghost of it . . . are often what our friends and intimates perceive as our most characteristic quality."[128] James then explained the significant role that this inner life plays in the constitution of a patient suffering from nervous disease. Its pathological components are apparently twofold, comprising both mental and physical aches and irritations:

> In the unhealthy-minded, apart from all sorts of old regrets, ambitions checked by shames and aspirations obstructed by timidities, it consists mainly of bodily discomforts not distinctly localized by the sufferer, but breeding a general self-mistrust and sense that things are

not as they should be with him. Half the thirst for alcohol that exists
in the world exists simply because alcohol acts as a temporary anaesthetic and effacer to all those morbid feelings that never ought to be
in a human being at all. In the healthy-minded, on the contrary, there
are no fears or shames to discover; and the sensations that pour in
from the organism only help to swell the general vital sense of secu-
rity and readiness for anything that may turn up.[129]

Benedikt's concept of an inner or buried life was peculiarly his own
and resisted easy classification alongside the competing theories of his
psychologically and neurologically minded contemporaries. With his
stress on the corporeal as well as the psychical components of this "second life," his ideas sat partially within the tradition of materialist authors
such as the French psychologist Theodule Ribot (1839–1916), who maintained that all psychological activity, indeed personality itself, could be
traced back to the workings of the physical body, while also sharing elements of the work of Ribot's compatriot and fellow psychologist Pierre
Janet.[130] Janet's research centered on the role of the psyche in creating
illness and mental distress, and James's interest in his ideas will be explored in more detail in chapter 3. In the first half of "The Gospel of
Relaxation," as discussed above, James dealt with the former idea and
explored how the condition and muscular tone of the physical body
influences the state of the mind. In the second half of his address, he
moved on to the latter topic, the toxic consequences of thoughts and
attitudes and their effect on a person's health.

In this section he insisted that, once again, Call had blazed a hygienic
trail that he advised everyone else to follow. He referred his readers to
her later book *As a Matter of Course*, which he described in a review article as a masterful work of mental hygiene that preaches "the gospel of
mental relaxation." He suggested that her book was let down only by
its "rhetorical ability"—her style tending more toward "sincerity and
plainness." Had her book only "a little more brilliancy," he insisted, "it
might well take a place with us similar to that which Feuchtersleben's
'Diätetik der Seele' has held in Germany, and go through edition after
edition."[131] In essence, he explained in his lecture, her gospel advocates
the "dropping of things from the mind, and not 'caring.'" This is good
advice, according to James, because it is supported by a psychological
"law" of "very deep and wide-spread importance in the conduct of our
lives." This is the principle that "*strong feeling about one's self tends to ar-
rest the free association of one's objective ideas and motor processes.*" This
mental state is exemplified, he continued, in the "mental disease called
melancholia."[132]

James described how a melancholic patient is "filled through and through with intensely painful emotion about himself." His mind is "fixed as if in a cramp" upon these feelings of guilt, doom, and fear. He pointed out that "all the books on insanity" report that the usual varied flow of such a man's thoughts have ceased, inhibited by his inward perseverance on the fact of his own desperate condition. There is, James claimed, a useful hygienic conclusion that can be drawn from this. If we wish our ideas and actions to be "copious, varied and effective," it is imperative that we form the correct mental habit, "of freeing them from the inhibitive influence of reflection upon them, of egoistic preoccupation about their results." "Just as a bicycle-chain may be too tight," he explained, "so may one's carefulness and conscientiousness be so tense as to hinder the running of one's mind." He went on to give the example of overpreparation for school examinations, insisting that "one ounce of good nervous tone in an examination is worth many pounds of anxious study for it in advance." "Worry," he surmised, "means always and invariably inhibition of associations and loss of effective power."[133]

Implicit in this section of James's "Gospel of Relaxation," and throughout his *Talks to Teachers*, is a definition of "mental health" that encompasses the notion of "efficiency" and not simply the absence of mental distress or irrationality. The "loss of effective power," decried by James in the passage above, is presented as a form of mental disorder, even though it is not a disease state, as such, and does not, itself, cause any painful symptoms. This association between health and work was ever present in James's discussion of his own maladies, and in a letter to a friend, in 1884, James stated that he had "long ago come to think that the right measure of a man's health is not how much comfort or discomfort he feels in the year, but how much work, through thick & thin, he manages to get through."[134] This linking of health with efficiency and the capacity for work was not peculiar to James's writings, however. Historians have long recognized that the "production of useful, creative labor" was the externally recognizable indicator of good health at this time.[135] This concern with optimum efficiency was a prominent feature of Ray's text, *Mental Hygiene*. There he wrote that

> philosophers lay down rules for the conduct of the understanding and sages mark out the path which leads to virtue and happiness. But something more is needed to enable the mind to act efficiently and to increase its capacity for labor and endurance.... The problem presented to every child of Adam is this,—having received certain powers of mind from nature, how are they to be managed so as to insure the greatest possible return?[136]

Ray's statement encapsulates the assumption that, in its fullest sense, good mental health, in America in the nineteenth century, was as much about maximizing the faculty for mental labor as diminishing the prevalence of pain and suffering. In his "Gospel of Relaxation" James pathologized worry, not only because of its inherently distressing quality but also because of its debilitating impact on the smooth operation of our mental machinery. These same ideas seem to underlie certain passages in his *Talks to Teachers*. On more than one occasion in these lectures he appeared to speak directly to the pupils themselves, rather than their teachers, and to offer them reassurance. He acknowledged that a mind that has a tendency to wander might be a liability when it comes to studying, and that a poor memory is a hindrance to gaining good examination marks, but he took pains to encourage those cursed in these ways and assure them that all is not lost:

> no one need deplore unduly the inferiority in himself of any one elementary faculty. . . . The total mental efficiency of a man is the result of the working together of all his faculties. . . . If any one of them do have the casting vote, it is more likely to be the strength of his desire and passion, the strength of interest he takes in what is proposed. Concentration, memory, reasoning power, inventiveness, excellence of the senses, are all subsidiary to this. . . . Our mind may enjoy but little comfort, may be restless and feel confused; but it may be extremely efficient all the same.[137]

On the topic of habits, he pointed out that, while bad habits can be harmful, good habits can be trusted to deliver us safely through adolescence and into adulthood with everything we require to become proficient workers.

> As we become permanent drunkards by so many separate drinks, so we become . . . authorities and experts in the practical and scientific spheres, by so many separate acts and hours of work. Let no youth have any anxiety about the upshot of his education. . . . If he keep faithfully busy each hour of the waking day, he may safely leave the final result to itself. He can with perfect certainty count on waking up some fine morning to find himself one of the competent ones of his generation.[138]

It would appear that, in offering these assorted words of comfort and reassurance, James hoped to dispel any anxious and thus pathologically inhibitive ideas that may be lurking in the minds of certain students. His encouragement was, in his eyes, a hygienic intervention. A decrease in anxiety both avoided the misery of melancholia and led to an increase in efficiency or, in other words, better health. The broad solution, he concluded in his "Gospel of Relaxation," was to stop worrying and to free "our American mental habit into something more indifferent and strong."[139]

In general, habits and health were at the heart of many of James's lectures and writings, particularly during the first two decades or so of his career. These two concepts were inextricably entangled in his thinking. The right habits were a crucial prerequisite for hygienic living, and according to the medical psychologists this was as true for habits of mind as of body. Indeed, according to James's Carpenterian ideas, these two types of habit were one and the same in any case. This insight was an integral element of James's chapter on habit in his *Principles of Psychology* and his *Talks to Teachers*. There he stressed the importance of acquiring the appropriate inventory of habits from an early age, while the nervous system was still in a relatively plastic state. Leaning on an agenda that was borrowed from Spencer, the model Jamesian education was one that produced healthy students; and a healthy student was one whose repertoire of habitual behaviors were those needed to survive in their environment.

Of foremost importance to James was the habit of self-control. He was convinced that the exercise of self-discipline was essential if a student was to have the ability to turn their attention away from distressing emotions and thoughts and toward more salubrious subjects. Unbridled emotions had long been depicted as poisonous to the body and mind in the mainstream hygienic literature as well as the more specialist texts produced by the alienist community, and this context is also an instructive one in which to set James's investment in his own controversial theory of the emotions that, as he explained in his lecture on "The Gospel of Relaxation," offered valuable insights into emotional control.

For James, the pursuit of health meant much more than freedom from disease or even the duty of every citizen to perform efficiently to the best of their ability. It was a genuine ideal, a blissful state to aspire to, in and of itself. His lecture notes indicate that he liked to end his public talks on alcohol with some "moral words," including a quote from the American journalist and folklorist Charles Godfrey Leland (1824–1903). He cited Leland's declaration that the feeling of intoxication induced by the world's finest champagne could not be compared with "such exhilaration as one feels from simple, sober, perfect health on a fine Indian-summer morning." He had tried both, "and of the two, the best excitement was that of my own bounding life-blood." James

was passionately devoted to the sentiment expressed by Leland and assured his audiences that the worship of health was "the coming ideal."
He may never have practiced as a doctor in the conventional sense, but
via his oratory James performed the role of a public physician, preaching the gospel of "perfect health" to society.[140]

CHAPTER 3

Religion and Regeneration

The Science of Organic Life

James is perhaps best known for one of the last texts he published: *Prag-matism* (1907). There he advocated an open-minded attitude toward the types of truth that should be taken seriously. He made the case that both the mystically minded prophets associated with religion and the aca-demic investigators of science are obliged to put their respective claims to the same practical test. Ultimately, he argued, "On pragmatic prin-ciples we cannot reject any hypothesis if consequences useful to life flow from it."[1] James's name has come to be synonymous with this theory of knowledge, and many have characterized him as a thinker torn by loyal-ties to both scientific and religious worldviews and determined, via his philosophy, to mediate between them.

In his lectures on *Pragmatism*, James's famous defense of the potential validity of religious truth was shorn of the context within which it was formed. Several commentators have implied, following the lead of his official biographer Perry, that this context comprised James's early years of "spiritual crisis." In effect they have constructed a direct intellectual bridge from the troubled young James in the late 1860s to the sixty-five-year-old author in 1907, which passes over the intervening forty years.[2] During this period, however, James's ideas were far from static; his un-derstandings of "God," "religion," and "science" and the relationship be-tween them morphed considerably, and these transformations were em-bedded, like so much of his thinking, in matters of a medical nature.

James's burgeoning interest in psychological means of recovery from both physical and mental maladies was central to these intellectual de-velopments. The notion that a patient's mind may, in one way or an-other, be the agent of their cure or "regeneration" to use one of James's preferred phrases, inspired a significant number of his later writings. These works also evinced a significant level of regard for the medical insights and experiences associated with a particular group of people: those James described, collectively, as "the Mystics." From the mid-1880s onward, he became increasingly respectful of those "wonder-mongers and magnetic physicians and seventh sons of seventh daughters," who keep and transmit "from generation to generation the traditions and practices of the occult."[3] James, however, had not always felt this way about mystical medicine. In his youth his loyalties lay exclusively with medical *science*.

As a young student James had high hopes that one day, albeit in the "dimmest future," the science of physiology would become "the light in which medicine is to work." In a review of the French physiologist Claude Bernard's *Rapport sur les progrès et la marche de la phsyiologie générale en France* (Report on the advance and progress of general physiology in France), which James published during his medical studies, he acknowledged that physiology as a discipline was in its merest infancy compared with others such as chemistry or physics.[4] He asserted, however, that

> the science of organic life has no qualitative inferiority to that of inorganic life: it is only far behind it in development. It is still embryonic,— the matter with which it deals is so endlessly complicated, that the end of the thread which is to guide the physiologist through its labyrinthine turnings has not yet come into his hand.[5]

He concluded that, although it may be a long wait, he foresaw a future where "*applied physiology* will . . . finally form, like applied chemistry, a body of knowledge having a certain roundness of its own" within which "the medical man may disport himself at greater ease than he does at present." He had faith that the current unwieldy mass of "undifferentiated lore" would be transformed, at the hands of the physiologists, into empirically grounded, bona fide laws of medical science.[6]

The battle cry of contemporary physiologists, such as Bernard, was that only via a careful elucidation of the elementary processes of life would it be possible to truly understand the workings of the human or animal organism. In his 1865 review of the English anatomist Thomas Huxley's *Lectures on the Elements of Comparative Anatomy*, James summarized the assumption that he believed underpinned such claims. They rest, he explained, on a "view of the phenomena of life which makes them result directly from the general laws of matter, rather than from the subordination of these laws to some principle of individuality, different in each case."[7] And, in a letter to a close friend, four years later in

1869, James invoked this same dichotomy of beliefs, one based on general laws and the other on individuality, to represent the difference between scientific and religious philosophies of life. He went on to note that "in developing, the Nature-lore [science] and the individual-fate-lore or religion have become so differentiated as to be antagonistic."[8]

According to James's definition contemporary "science" allowed only for impersonal, universal laws of matter and anyone who believed, on the contrary, that "events may happen for the sake of their personal significance" was partaking of a religious worldview. In later texts James conceded to his readers that his own definition of "religion" may be at odds with theirs in that it encompassed what others might refer to as "moralistic" rather than religious schemes. In his earlier writings, however, no such differentiation was made, and James, by his own admission, tended to associate all "religious thinking, ethical thinking, poetical thinking, teleological, emotional, sentimental thinking" together. He collectively labeled them "the personal view of life to distinguish it from the impersonal and mechanical, and the romantic view of life to distinguish it from the rationalistic view." As he later put it, "for mechanical rationalism, personality is an insubstantial illusion."[9]

In his Huxley review James appeared to go along with the author's working hypothesis that "the phenomena of life" do indeed follow the natural laws laid down by science, rather than bend themselves to the "religious" principle of individuality. He conceded that such a standpoint seems to appear unremittingly atheistic but suggested that perhaps it might, nevertheless,

> serve an excellent purpose, and in the end, by introducing order into the Natural, prove to be a necessary step in the way to a larger, purer view of the Supernatural . . . it will do away at any rate with that eternal muddling together of Natural and Supernatural which has prevailed hitherto. God will no longer be made to appear as on a level with Nature and acting as a mere rival to her forces.[10]

May it not be, James surmised, that "finding Nature a great closed sack, as it were, . . . without any *partial* inlets to the Supernatural, without any *occasional* Ends *within* her bosom, we shall be driven to look for final causes on some deeper plane underlying the whole of Nature at once, and there shall find them?"[11] At this stage in his life, it would seem that James did not expect to find any lawless gaps in nature, miraculous interventions that supersede the order of natural processes. Nor did he see such a state of affairs as inherently atheistic. He conceived of the possible existence of a supernatural realm that would make moral sense of, but not physically interfere with, the day-to-day running of the natural world.[12] When it came to comprehending and manipulating this natural world, including the functioning and malfunctioning of the human body, it was scientific wisdom that James looked to for answers.

The young James's faith in medical science was exemplified in a review, which he published in 1868, about a book by a French physician Ambrose Liébeault (1823–1904) called *Du sommeil et des états analogues* (Of sleep and analogous states). When it was published the book attracted very little attention, and James was one of the few who noticed it. It was not until the mid-1880s that Liébeault's ideas garnered more widespread interest and enthusiasm after being taken up by the French physician Hippolyte Bernheim (1840–1919).[13] James gave his review the title "Moral Medication," and he explained that the purpose of Liébeault's treatise was to show the "essential identity" of "artificial or 'magnetic' sleep, the so-called hypnotic state of Braid," with "ordinary slumber" and to "advocate the application of it to the treatment of various forms of disease."[14]

James's use of the term "'magnetic' sleep" was a reference to the, by then notorious, healing practices of animal magnetism that were developed originally in eighteenth-century France, by Franz Anton Mesmer (1734–1815). Mesmer's early cures utilized magnets, and he claimed to have discovered a vital life force, an invisible fluid or energy called "animal magnetism." Under his disciple Amand-Marie-Jacques de Chastenet, Marquis de Puységur (1751–1825) "magnetizing," or "mesmerizing," a patient came to be synonymous with putting them into an unusual sleeplike state of consciousness or trance. Puységur also insisted that the success of animal magnetism lay "not in the action of one body upon another, but in the action of thought upon the vital principle of the body."[15]

These mental healing techniques were eventually introduced into America by traveling lecturers, in the early nineteenth century, and, by the 1860s, there were significant numbers of these magnetic lay practitioners offering their services as an alternative to treatment by the professional physicians. Liébeault and the Scottish surgeon James Braid (1795–1860), who coined the term "hypnosis," were among the earliest members of the medical professions in their respective countries to advocate the merits of these controversial mesmeric practices to their colleagues.[16]

In his review James applauded Liébeault for his attempts to construct a physiological explanation, albeit, he acknowledged, a crude one, of the "apparent miracles" associated with animal magnetism. He recounted Liébeault's theory that during the state of artificial sleep, or somnambulism, many points of the somnambulist's sensibility are "paralyzed and incapable of vibration." In effect, the sleeper's entire nervous system attends to only a very limited number of incoming sensations and, as a consequence, becomes intensely concentrated on this narrow range of impressions. This exalted level of extreme mental focus was, James related, "probably the cause of the apparent miracles which somnambulists perform," including, for example, the case of the subject who had "distinguished infallibly by his tongue, among six glasses of water . . . one of which had been warmed by some one's finger-tips being held over it for two minutes." James related the author's conclusion, namely, that this faculty of concentrated attention may be put to good use and "enlisted in the service of therapeutics. As the mind calls forth disease in hypochondriacs and hysterics, so it should be employed to banish it." This procedure, he explained, involves the doctor issuing curative verbal "suggestions" to the patient, who has been placed in a state of artificial sleep: "Of course, the principle in suggestion is to divert the attention, to displace it from morbid symptoms, and to concentrate it upon [bodily] functions which are torpid [sluggish]."[17]

The review, which was written while James was a medical student, is notable both for his low opinion of lay healers and the concomitant high esteem in which he held the medical profession. He prefaced his own summary of Liébeault's clinical results with the comment that they comprise "a contribution to moral medicine of more value than the records of animal magnetism." He ended, moreover, with a plea to his colleagues:

> It seems high time that a realm of phenomena which have played a prominent part in human history from time immemorial should be rescued from the hands of uncritical enthusiasm and charlatanry, and conquered for science; . . . educated medical men, . . . [should] shake off the discreditable shyness which has hitherto characterized them, and walk boldly in to take possession.[18]

At this point in time, James was convinced that medical miracles were simply events for which the correct physiological laws had yet to be determined, and he assumed that the project of science paved the sole legitimate pathway to healing truth. During the 1880s, however, James's attitude began to change, and his personal correspondence bore witness to the first signs of dissatisfaction with the teachings of the medical profession and disillusionment with the scientific laws of health. In August of 1885 he wrote to his friend and fellow philosopher George Croom Robertson (1842–92) expressing his exasperation:

> After a certain amount of experimenting on diets etc, I have concluded to throw regimen to the dogs, and live according to the temp-

tations of the world, flesh & devil. Words cannot express my contempt for much of the medical legislation that is current, based as it is
on a theoretic surmise, grafted on an insufficiently observed fact, generalized by pedantry, promulgated by love of dominion, and adopted
by credulity as the rule of life. Let every man be his own observer and
believe in no results that are not obvious; and let him be ready for the
angel with the scythe whenever it comes.[19]

James's contempt for the contemporary medical laws and the lawmakers is obvious, but, equally, the bitterness of his frustration implies
the belief and trust that he had once invested in them. In this respect
the timing of this letter may have been significant; the "angel with the
scythe" had recently taken the youngest member of James and Alice's
family. After struggling with whooping cough and associated medical
complications, their third son, Herman, had died a month earlier, in
July 1885, at the age of seventeen months. It is open to speculation as to
whether medical science's failure to prevent the suffering and death of
his son had contributed to James's loss of faith in physiological wisdom.
In any case, a letter written a few months earlier indicates that his curiosity had already been piqued by the medical implications of the new
field of "psychical research."

The Wonder-Mongers

In February of 1885, James wrote to Frederic Myers (1843–1901), a key
figure in the Society for Psychical Research in England. The remit of
the society, which was founded in London in 1882, was to investigate,
without prejudice, spiritualist mediumship, thought transference, and
other such phenomena. Myers had recently authored two papers dealing
with what was known as "automatic writing." Automatic or involuntary
writing was the name given to the phenomenon whereby someone produced written material without a conscious volition to do so and while
apparently unaware of its content. It was often popularly attributed to
the actions of a spirit that had temporarily taken possession of the subject's body. Myers, however, had proffered an alternative explanation.
His suggestion was that at least "*some* of the effects which Spiritualists
ascribe to spirits are referable to the unconscious action of the writer's
own mind." His theory relied on "the conception of an obscure ocean
of sub-conscious mental action;—hidden waves whose shifting sum-
mits rise for a moment into our view and disappear."[20] James's immediate reaction to Myers's hypothesis was to connect it with the world of
mental disease:

What you have written seems to me to cast a new light on those cases of mania in which the person seems doubled or transformed. But how to conceive of this "unconscious" other fellow in us, I confess I don't yet know. I imagine that owing to your labours and those of others a greater work in the way of remodelling scientific prejudice is going on than appears on the surface.[21]

In the same letter, James speculated about whether Myers's theories could also explain the research he had previously carried out into the sensations that many patients continue to perceive as originating from within a lost limb.[22]

Later that year, in the autumn after the death of their son, James and his wife were introduced to Leonora Piper (1859–1950), a spiritualist medium who very quickly impressed James. Piper became James's experimental subject of choice regarding his own contributions to psychical research. He was convinced that "she knows things in her trances which she cannot possibly have heard in her waking state." In an account of these initial investigative trance sessions, which he wrote a few years later, James described how she was able to answer questions about distant family members and other private topics that she could not have learned about via any ordinary means. Subsequent sittings and a personal acquaintance with her led him to the confident belief that she had "supernormal powers."[23] These powers, moreover, extended into the realm of spiritual healing.

During the first ten years or so of her career as a medium, Piper's "spirit control" purported to be the spirit of a dead French doctor.[24] A letter from the psychical researcher Richard Hodgson (1855–1905) implies that much of Piper's time, during this early period, was taken up by consultations with the sick. In her mediumistic trances her whole persona would change, and she spoke in the "deep gruff voice" of a Dr. Phinuit.[25] During the late 1880s both James and his wife appear to have sought Dr. Phinuit's help with health matters. In October of 1889, for example, James wrote to Alice and mentioned that he was keeping well as a result of the advice given him by Dr. Phinuit. In another letter he wondered aloud whether "Mrs Piper's pill" might have given him a headache, intimating that the dead doctor's spirit went so far as to prescribe medication. In some instances Phinuit's medical expertise appeared to James to have been vindicated. In one letter James reported that his brother-in-law, William Mackintire Salter, was "already better for Mrs P. How queer it would be if that cured him!"[26] Piper's daughter, Alta Piper, stated that the series of her mother's sittings supervised by the psychical researcher Oliver Lodge included many recorded accounts of "Phinuit's

3.1 James sitting with Mrs. Walden at a séance. They appear to be holding a pair of "spirit slates" between them, which were used by spiritualist mediums to communicate messages from the dead. James was a founder member of the American Society for Psychical Research, which was established in 1884, and chair of the Committee on Hypnotism. (Letters to William James from various correspondents and photograph album [MS Am 1092 (1185), 1, Box: 12], Houghton Library, Harvard University.)

diagnoses of physical ailments with, in some instances, the prescriptions which he recommended and which, when tried, proved efficacious."[27]

In this way, Piper plied two different spiritualist trades; some of her time was spent facilitating conversation between the living and their deceased friends and family, and on other occasions she offered a line in supernatural medical advice. This duality of vocation appears to have been commonplace within mediumistic circles. She first discovered her own purported spiritual powers, in 1882 at the age of twenty-five, after visiting a blind clairvoyant healer who was "then attracting considerable attention by his remarkable medical diagnoses and subsequent cures." Her father had urged her to visit the healer in the hope that he could help his daughter, who was suffering from the effects of an accident that had taken place many years earlier. During the session, the clairvoyant diagnosed Piper's troubles (accurately it was later ascertained), and Piper briefly lost consciousness. Her daughter recorded that her mother was "disturbed and greatly puzzled by her experience." She was subsequently persuaded, however, to attend one of the regular meetings or "circles" that the clairvoyant held for the purpose of effecting cures and developing latent mediumship. During this circle Piper delivered her first spirit message, which was allegedly from the dead son of one of the other attendees.[28]

Historically, the spiritualist community and the tradition of lay healing were closely linked. One of James's contemporaries, fellow psychical researcher Frank Podmore (1856–1910), wrote a historical account, in 1909, entitled *Mesmerism and Christian Science: A Short History of Mental Healing*. In this text he explained that both groups shared a common root in the practices of mesmerism and, more specifically, the associated trance state. Podmore clarified that the trance state came to assume two different therapeutic roles. It was treated primarily as a curative experience for the patient themselves but also secondarily as a means of "diagnosing and prescribing for the ailments of others." However, from the 1840s onward, he asserted, "a large number of those who had hitherto practised Animal Magnetism or Mesmerism, whether in America or Europe, were sooner or later absorbed into the ranks of the Spiritualists." For these practitioners, their "chief interest" in "the trance" lay "no longer in its therapeutic possibilities, but in the promise which it holds forth of spiritual revelations." The mind-curers, or late nineteenth-century mental healers, comprised those "whose interest in healing was greater than their love of the marvellous."[29]

In light of the content and chronology of James's correspondence it seems likely that it was Piper who first convinced him that these lay healers may have genuine medicinal powers worthy of serious study. A few months after he first met her, he wrote to a family friend and unorthodox physician, James John Garth Wilkinson, about the work of the psychical researchers. For the first time James discussed this research in terms of what it might mean for the claims of the "magnetic physicians."

He opened the letter with a critique of Wilkinson's recent work, *The Greater Origins and Issues of Life and Death*.[30] Wilkinson was well known for his opposition to the medical profession and the authority they assumed. The front cover of a pamphlet edited by him, for example, featured the quote: "The science of medicine is founded on conjecture, and improved by murder."[31] James expressed his concern that Wilkinson went too far in his reaction against the "incredible shallowness" of the materialist assumptions adopted by those medical researchers who allied themselves with the "sect" of "'scientism.'" Referencing the work of Louis Pasteur et al. on the microbial causes of disease, he articulated his contrary position that, in one respect, appears to have changed little since his review of Huxley's lectures twenty years earlier:

> You go too far and misprize the fact that this world, whatever it may be besides, is at any rate also a machine, and a place in which the higher things must get the lower ones into their service and "obey" their laws, before they can possibly *be*. I don't see why microbes *may* not be, from the point of view of practice, the cause of any disease, and the study of what is lethal thereto, the only study of cure. If it were so, it would only be one more item in the materialism of things. Why isn't the world's body all mechanical, and yet *all* the mechanism for the sake of a spiritual end? Instead of part being mechanism & part end, as ordinarily supposed?[32]

James, although newly curious about the phenomena reported by the psychical researchers, revealed himself to be still firmly wedded to the idea that such phenomena can and should be disciplined by the natural laws of matter. Whereas the ultimate *meaning* of the world's events may be connected with a spiritual purpose, these happenings must also be accounted for, he implied, by the mechanistic principles of science. He did, however, express his support for Wilkinson's general standpoint, namely, his opposition to the "sottish sect of 'scientists' so far as it is a sect—which of course it is on an enormous scale." James continued: "When I think of the resolute ignorance and conceited barbarism of these fellows who turn their hindquarters to the sky and tell you that is the only scientific attitude, I almost foam at the mouth." But, James avowed, he envisaged an optimistic future for scientific inquiry, one in which both Wilkinson's spiritualistic approach to medicine and the more materialist doctrines would play a role:

> Your book therefore I simply take as one way of taking in the sum of things, destined later to be completed by fusion with what now seems to be its alternative. It shows what through all this "Psychical Research" I am coming to believe as I never did before, that the fulness of truth is not given to any one type of mind. I have hitherto felt as if the wonder-mongers and magnetic physicians and seventh sons of seventh daughters and those who gravitated towards them by mental affinity were a sort of intellectual vermin. I now begin to believe that that type of mind takes hold of a range of truths to which the other type is stone blind. The consequence is that I am all at sea, with my old compass lost, and no new one, and the stars invisible through

the fog. But it is exhilarating to have things suddenly enlarge their possibilities — at any rate.[33]

It is clear from his letter that James's map of the medical landscape had been recently and radically redrawn. His newfound respect for the spiritualists and magnetic physicians had brought with it a glimpse of a fresh field of vision: the possibility that useful medical knowledge may originate outside of the scientific study of life and that this lay healing community possessed a unique and valuable perspective that would prove indispensable in the accumulation of certain "truths." On a more practical level, however, within a year of his communication to Wilkinson, James began to explore the therapeutic services of the "wonder-mongers" for himself. In the winter of 1887 he asked his friend George Bucknam Dorr to introduce him to a local "mind-cure doctress."

James's tendency to refer to contemporary lay healers collectively as "mind-curers" was somewhat misleading as they belonged to an array of different groups and organizations. The various schools of "faith healing," as he also described them, included Christian Science and New Thought. Despite fiercely antagonistic interrelations, between these two groups especially, James was of the opinion that "their agreements are so profound that their differences may be neglected."[34] These points of agreement centered on one common doctrine: mental therapies alone were sufficient to cure all disease.[35]

The origins of the American mind-cure movement can be traced to the activities of an eccentric clock maker from New Hampshire, called Phineas Parkhurst Quimby (1802–66). During his lifetime he developed and practiced his own mental therapeutics, claiming to have treated twelve thousand men, women, and children. Inspired originally by demonstrations of mesmerism, Quimby's central tenet was that all diseases are mentally induced. Among the thousands of patients he treated, four of them, Warren Felt Evans, Mary Baker Eddy, and Julius and Annetta Dresser, went on to found their own schools of mental healing after his death.

During the late nineteenth century the mind-cure movement spread rapidly across the country. Proponents of the cause wrote pamphlets and published periodicals offering advice and guidance to their readers on how to cure themselves at home. But in the early days at least, Boston, just a few miles from James's home in Cambridge, Massachusetts, was the center of the mental healing community. By the mid-1880s Mary Baker Eddy had founded her church of Christian Science and moved there to establish a training school for those wishing to practice her healing methods. Similarly, Warren Felt Evans, the Dressers, and Edward J. Arens had all established competing institutions in the area. A local newspaper reported that, in addition to these four major schools, "there are about a dozen others who practice the mind cure as a profession, and who teach classes of young and old the methods of curing."[36]

Correspondence with the German philosopher and psychologist Carl Stumpf (1848–1936) suggests that it was trouble sleeping that prompted James to make his first personal experiment with mind cure. In a letter to his cousin Kitty Prince, he wrote that he had made ten visits to a "mind-cure doctress," but added, "I cannot see that the mind cure has done me any positive good, though I shall go twice more, having resolved to give the good woman at least a dozen sittings, for fair trials sake. She has done wonders for some of my friends."[37]

The identity of the mind-curist is not revealed directly in this corre-spondence, but cross-referencing James's letters from this period indi-cates that she was Annetta Seabury Dresser (1843–1935), one of the key figures in the world of nineteenth-century mental healing.[38] (Her first son, Horatio Dresser, would later study for his PhD at Harvard under William James and write various books on mental healing and its his-tory.) The Dressers moved to Boston and began their healing work there in 1882, so their circumstances are compatible with the timing of James's visits.[39] As to what went on during these therapeutic appointments, a letter to his sister, Alice James, provides more details:

> I sit down beside her and presently drop asleep, whilst she disentan-gles the snarls out of my mind. She says she never saw a mind with so many, so agitated, so restless etc. She said my *eyes* mentally speaking, kept revolving, like wheels in front of each other and in front of my face and it was 4 or 5 sittings ere she could get them *fixed*. I am now, *unconsciously to myself*, much better than when I first went, etc. . . . Meanwhile what boots it to be made unconsciously better, yet all the while consciously to lie awake o' nights as I still do?[40]

It is possible that the sleep enjoyed by James during his sessions might have been "mesmeric sleep," which formed part of the therapeutic arsenal of many mental healers. In this case, it seems a prosaic explana-tion is more likely, since in his letters James mentioned that he deferred his regular afternoon nap so that he could take it during his mind-cure appointments.[41] His healer's identity is also relevant with respect to this matter. Annetta Dresser wrote a book about the healing philosophy and techniques that she had learned as a patient and acquaintance of Quimby, and since she is nothing but admiring of Quimby's procedures in her book, it seems reasonable to assume that her account is a fair rep-resentation of the methods she herself employed.

In *The Philosophy of P. P. Quimby*, Dresser stated explicitly that it was only in his early years as a healer that Quimby had experimented with mesmerism and placing subjects into "a state known as the mesmeric sleep." This she described as "more properly a peculiar condition of mind and body, in which the natural senses would or would not operate at the will of Mr Quimby." Later on, she insisted, he became disillusioned by such methods and found no use for them in his work.[42] Apparently Quimby's and, we may assume, Dresser's healing relied solely on their ability to change their patients' apprehension of their illness.[43]

According to Dresser, Quimby's fundamental insight was that disease is "an error of the mind, and not a real thing." She explained that "as one can suffer in a dream all that is possible to suffer in a waking state, so Mr Quimby averred that the same condition of mind might operate on the body in the form of a disease, and still be no more of a reality than was the dream." In Quimby's own words, all doctors "admit disease as an independent enemy of mankind . . . [whereas] I deny disease as a truth, but admit it as a deception, started like all other stories without any foundation, and handed down from generation to generation till the people believe it."[44]

According to Quimby, the act of healing involved sitting with the patient and "feeling" what their problem was. He believed that a sick body was the product of a mind that was laboring under the erroneous notion of disease. Furthermore, "when the mind or thought is formed into an idea, the idea throws off an odor: this contains the cause and effect." This "odor" or mental atmosphere, which emanated from the patient, was sufficient to tell Quimby all he needed to know about their illness. Dresser recounted how, having ascertained what false idea or "superstition" they labored under, he would converse with them and explain the cause of their symptoms. He would "thus change the mind of the patient, and disabuse it of its errors and establish the truth in its place, which, if done, was the cure." Sometimes, however, audible words were not necessary to affect a cure. The sensing of the odor could also be done in the absence of the patient, "at a distance of many miles," and the cure could also be achieved at a distance. This was possible, Quimby had concluded, because he came to see "that the senses could act independent of the body, and that the five natural senses, . . . embraced but a small part of man's perceptions. . . . Man then, . . . is capable of hearing, seeing, smelling and communicating thoughts and feelings without the aid of matter."[45]

This description, of a nonvocalized process of cure, is compatible with James's own description of Dresser's actions and would explain why she seemed convinced she had healed him, despite him having been asleep during the sessions and unconscious of any change. Although

James was skeptical at first, a letter to his brother, written a month after the letter to his sister, revealed a more circumspect attitude toward his experience: "We are both well, I in particular, so much so that I wonder if it be not due to the Mind Cure in which I have been dabbling. I have had a new lease of youth of late, and enjoy going out and sitting up late."[46] Two weeks later, however, his final word on the subject, to another correspondent, seems to have been that Dresser was "inert" in his case but was a "good woman."[47]

Despite James's initial reservations about the efficacy of mind cure in his own case, there is evidence that he remained far from dismissive of the movement in general. During his lifetime James consulted a range of different mental healers for at least ten separate courses of treatment. Each of these consisted of between ten and twenty individual sessions, and, on at least one occasion, he credited the healer in question with significant success.[48] In a letter to his wife, in the year after his visits to Dresser, he related how he had "defended those who [are] interested in the mind cure etc" to social acquaintances of theirs.[49] It seems likely that it was the favorable accounts of close friends and family that prompted his support. There is evidence, for example, that Alice also experimented with their healing methods. James noted on one occasion that she seemed to be averse to "letting rip" with her feelings about certain individuals and ventured the suggestion that perhaps it might be "the Mind-cure that makes you afraid of being harsh in your talk." (The doctrines of mind cure taught that in addition to thinking and speaking positively about yourself and your own health, you should also extend this attitude toward others.)[50] In another letter that year he mentioned "a magnetic healer," called Mrs. Wetherbee, of whom several members of his wife's family had given good accounts.[51]

In short, it would seem that lay mental healing was, by 1888, an established and popular recourse for troubled members of James's extended family, and some of his friends, and that James trusted these accounts despite his own, less fruitful, first experience. There is, however, no indication as to how James conceived of the phenomenon in a psychological sense or that he had any idea of how to account for its success. A year later though, James appears to have experienced an epiphany in the form of French psychologist Pierre Janet's doctorate of philosophy thesis *L'Automatisme psychologique* (Psychological automatism).[52] In October 1889 James wrote to Dorr, the man who had first introduced him to Dresser, telling him that he had been reading Janet's recently published thesis, "which tells more queer new facts about the secondary or subconscious self than any work yet published." He insisted that Dorr "must read it" since "all Dresserism [mind cure] is surely connected therewith."[53]

The Hidden Self

Janet had investigated several cases of hysterical trance or somnambulism while working as a professor at the Lycée in Le Havre, and he continued his research in this field under the tutelage of the neurologist Jean-
Martin Charcot (1825–93), at the Salpêtrière Hospital in Paris. In his
thesis Janet outlined a theory of mental dissociation to account for the
condition of hysteria. He proposed that elements of the patients' original
personality had become separated and developed autonomously in subconscious strata of their mind, alongside their ordinary consciousness.
These subconscious secondary personalities were, he alleged, responsible for the various symptoms displayed by the hysterics he had studied. Janet had previously written several articles about his investigations
into the hypnotic state and the disease of hysteria, but none of these had
included his account of how he cured a young female patient named Marie.[54] It was this particular case study, which appeared in print for the first
time in his doctorate, that captured James's imagination.

In 1890 James published a summary of Janet's thesis in *Scribner's Magazine*, under the title "The Hidden Self." In this article he expounded,
at some length, Janet's observations and theories about "what has long
been vaguely talked about as unconscious mental life." James, while acknowledging the work of other authors, such as Frederic Myers's experimental partner Edmund Gurney (1847–88) and the researchers Hippolyte Bernheim and Alfred Binet (1857–1911), insisted that Janet had
taken these preexisting inquiries into the "submerged consciousness" to
a new level. He proclaimed that the "simultaneous coexistence of the different personages into which one human being may be split is the *great*
thesis of M. Janet's book."[55]

Later in his essay James made the case that "the really important part"
of Janet's investigations was their "possible application to the relief of
human misery." James stated explicitly that it was Janet's extension of his
theoretical insights into the practical world of medical therapeutics that
was "more deeply suggestive to [him] than anything in Janet's book." To
illustrate this point, he gave an account of his cure of the nineteen-year-
old patient named Marie. Marie had entered the hospital in Le Havre
where Janet worked suffering from all sorts of different symptoms:

> At first, M. Janet, divining no particular psychological factor in the
> case, took little interest in the patient, . . . [who] had all the usual
> courses of treatment applied, including water-cure and ordinary hyp-
> notic suggestions, without the slightest good effect . . . M. Janet [did
> then] try to throw her into a deeper trance, so as to get, if possible,

some knowledge of her remoter psychologic antecedents, and of the original causes of the disease, of which, in the waking state and in ordinary hypnotism, she could give no definite account.[56]

James went on to relate how these experiments had succeeded beyond Janet's expectations when, in her deeper trance, Marie's subconscious personality was able to reveal all the psychological causes of her various symptoms:

> Her periodical chill, fever, and delirium were due to a foolish immersion of herself in cold water at the age of thirteen. The chill, fever, etc., were consequences which then ensued; and now, years later, the experience then stamped in upon the brain for the first time was *repeating itself* at regular intervals in the form of an hallucination undergone by the sub-conscious self, and of which the primary personality only experienced the outer results.[57]

Similar autobiographical narratives were supplied by the subconscious self to explain Marie's frequent terror attacks and localized anesthesias. Furthermore, by dint of inventive manipulation of these pathological memories, Janet managed to convince this secondary self to adopt a series of alternative, happier endings that resulted in the disappearance of all her troublesome symptoms. In James's words, Marie's ill health was the result of some "perverse buried fragment of consciousness obstinately nourishing its narrow memory or delusion, and thereby inhibiting the normal flow of life." These "senseless hallucinations" of her subconscious personality had become "stereotyped and habitual," hence the need for Janet to take this personality back in time to refashion the very memories themselves.[58]

This temporal dimension to Janet's treatment, which located and targeted the psychological cause of Marie's symptoms at a specific moment in the past, marked it out from the earlier hypnotherapeutic methods and represented a new opportunity for the field of suggestive cures.[59] Janet's sophisticated methodology did not appear to be the only significant aspect of his work from James's perspective, however. In his "Hidden Self" essay James implied that the most important element was the light he had shed on the role of the subconscious self in the etiology of disease. At the end of his essay he acknowledged that Janet associated the existence of secondary personalities exclusively with the disease of hysteria, whereas he believed that the French researcher's findings may prove to be relevant to a much larger section of the patient population. James declared: "Who knows how many pathological states (not simply nervous and functional ones, but organic ones too) may be due to the existence of some [morbid] . . . buried fragment of consciousness."[60]

In many respects Janet's work represented a continuation of older explanations of the psychogenesis of disease. The possibility that the mind may induce specific illnesses in the body had already been the subject of a certain amount of speculation by other authors. In the literature on medical hypnosis from the late 1880s, there was limited discussion as to how the mind may create disease, but there were references to the perils of pathological "autosuggestion." Just as the power of suggestion could be harnessed, beneficially, by the hypnotherapist, so it might also, inadvertently, be initiated by the patient themselves with adverse results. In essence, it was thought that if a person allowed a pernicious fear of illness to take hold of their mind it was quite possible that these morbid thoughts could lead to actual disease. These harmful ideas might originate, for example, from unwitting friends and family members who had commented to the person in question that their appearance suggested ill health. In other cases, it was proposed that the patient might have become concerned by a small change in their bodily sensations.[61]

The English physician and psychical researcher Charles Lloyd Tuckey (1854–1925), who wrote *Psycho-therapeutics, or Treatment by Hypnotism and Suggestion*, linked such notions of pathological autosuggestion with earlier discussions about hypochondriacal illness. He cited the theories of the English alienist Forbes Winslow who maintained that there was a psychophysical explanation for these types of occurrence. He recalled Winslow's claim that an organ in the body can become diseased as a result of "the faculty of attention being for a lengthened period concentrated on [its] . . . action." This intensity of mental attention, it was proposed, led to an increased blood flow to the area in question that, in turn, put a strain on the tissues and produced clinical consequences.[62]

Janet's account of Marie's troubles spoke to this preexisting literature on the mechanism of "autosuggestion," in that Marie's subconscious self brought about genuine physiological symptoms via its "obstinate" reliving and reexperiencing of morbid events and illnesses from the past. It had dwelled incessantly on the production of ill effects in her body, much as a hypochondriacal patient would, and, as a consequence, had induced these very results.[63]

There was one crucial difference with regard to Janet's paradigm, however. Marie could not control, and thus could not be held responsible for, her autosuggested illnesses, because she knew nothing of them. Nor could she be accused of "shamming," of making up or exaggerating her symptoms. They did not belong to her; they belonged to her secondary self: a self that may have shared her body but which had, quite literally, a mind of its own. This facet of Janet's theories, which turned on issues of volition and legitimacy, was of profound significance for James. In his "Hidden Self" essay he alluded to these implications of Janet's work. There he depicted the new research into subconscious states of mind as an opportunity to overthrow conventional medical attitudes toward "moral medication" and the psychogenesis of disease.

> The ordinary medical man, if he believes the facts at all, dismisses them from his attention with the cut-and-dried remark that they are "only effects of the imagination." It is the great merit of these French investigators, and of Messrs. Myers, Gurney, and the "psychical researchers," that they are for the first time trying to read some sort of definite meaning into this vaguest of phrases. Little by little the meaning will grow more precise.[64]

The words of the English physician Daniel Hack Tuke (1827–95) provide a revealing context for James's comments about the "ordinary medical man." In the second edition of his text *Illustrations of the Influence of the Mind upon the Body in Health and Disease*, published in 1884, he described the prevailing attitude of the medical profession toward mental therapeutics. Tuke explained that their current standpoint cast aspersions on both patients and practitioners who engage in such "heterodox," or in other words nonsomatic, methods of healing:

> Because effects are produced and cures performed by means of a mental condition called the Imagination, it is constantly assumed that these results are imaginary, in other words, that they are "all fancy." . . . It is generally implied that these phenomena are of a merely functional, subjective character, more or less dependent on the state of mind, more especially the Will, and that a change of mental condition has been naturally followed by a change in the phenomena, although apparently physical. . . . This is what the orthodox medical practitioner means, as he complacently smiles, or is indignant, when the success of his heterodox rival is dinned into his ears, and he asserts that it was all the effect of the Imagination.[65]

In other words, any symptoms that may be cured by mental means were never truly symptoms at all and were merely the "fancy" of the patient. As such, these illnesses are entirely subjective and may come and go at the will of the sufferer, making their cure a feat of no real medical significance. From James's perspective, the researches of Janet et al. subverted these pervasive and long-standing assumptions.[66] The existence of subconscious strata of mind exploded the old medical prejudices and legitimated the notion that psychological causes could lead to genuine disease.[67] This new field of study offered redemption for the "nervous invalid," who was no longer the agent of their own suffering, and it was a nail in the coffin of the doctrine of medical materialism.

James was so persuaded by these theories about the psychogenesis of disease that he feared the public health implications. In 1894 he sent a missive to the editor of the *Nation* entitled "The Medical Advertisement Abomination," where he went so far as to wage war on the advertisers of medical cures, on the basis that their ubiquitous portrayals of illness and morbidity were deliberately poisoning the population. The authors of such advertisements should be treated, he concluded, as "public enemies and have no mercy shown":

> The essays, the anecdotes, the elaborate accounts of symptoms, the portraits of sufferers, are a direct attack of the most efficacious sort upon the public health. They are meant to be such an attack; they have no other conceivable aim or intention than to produce panic, to beget in susceptible readers, by their incessant repetition, the fixed idea and apprehension of disease. Such fixed ideas and apprehensions are among the most potent morbific agents known; and the amount of hypochondriacal misery and actual disease caused by these suggestions would certainly, if it could be measured, prove almost incredibly great.[68]

On a more optimistic note, however, James welcomed the new explanatory possibilities that this line of medical research opened up for mental therapeutics. According to James, the organizing principle of a capricious, subconscious personality made sense of what had previously seemed anomalous about various forms of healing by suggestion. Two years after his essay on "The Hidden Self," in 1892, James reviewed Hans Schmidkunz's *Psychologie der Suggestion* (Psychology of Suggestion). There James related these curious incongruities:

> The truth is, that all the more *distinctive* phenomena of suggestion run dead against the *ordinary* laws of belief and practice. Why does a man who holds his own obstinately against every opponent in all the ordinary arguments of life submit to everything you tell him after he has made himself passive for a minute and you have performed a little hocus-pocus of passes over his face?[69]

The "philosophy of the subject" lay, James surmised, in the direction of "Messrs. Myers's, Binet's and Janet's researches into the different *strata*

of which personality consists." He made the case that hypnotic suggestions have consequences very different from those administered to the
patient in the waking state because they make contact with a different
level of their mind: "The *hypnotic stratum* must be thrown uppermost,
as in the trance, or in some way the suggestion must penetrate to it and
tap it, or there will be no effect."[70] A successful "suggestion," he proposed, is a suggestion that reaches the subject's subliminal region of
consciousness.

As James's letter to Dorr indicated, Janet's linking of the subconscious
self with the etiology of disease also facilitated, for James, an explanation of the cures performed by the lay mental healers, or mind-curers.
In "The Hidden Self" he asserted that these practitioners "unquestionably get, by widely different methods," results "no less remarkable" than
Janet's cure of Marie.[71] Their methods were indeed "widely different" in
that they frequently took place, as discussed earlier, in silence and even
at a distance. In 1894 James reviewed a book by Leander Whipple, the
founder of the *Metaphysical Magazine* and a practicing mental healer.[72]
In his book *The Philosophy of Mental Healing*, Whipple informed his
readers about one particular case where the healer and the healed were
located in different states, nearly a thousand miles apart.[73] In his review James criticized Whipple for providing insufficient information
about the techniques employed by the mind-curers to rid patients of
their pathological "subconscious fixed ideas." Whipple had, he pointed
out, given "no detailed account of the practical method by which the
fixed ideas are to be pulverized away." James then went on to give his
own preferred explanation. He suggested that "metaphysical healing" involves "something like the telepathic action of one subliminal self upon
another."[74] James's deployment of the term "telepathy" is significant in
that it was first coined, twelve years earlier, by Myers.[75]

In his earlier articles on automatic writing, Myers had endeavored to
explain as many so-called "spiritualist" phenomena as possible without
recourse to the idea of any kind of spirit entity independent of the human body. His hypothesis of the existence of telepathy, or "the communication of any kind from one mind to another independently of the recognized channels of sense," comprised one such attempt.[76] He applied
his telepathy theory to cases where someone had seen an apparition of
a relative around the time of their death, for example. In such instances,
he theorized, it was not their spirit that had come to visit them; instead
the dying relative had communicated the idea of their presence in that
place to the mind of the observer who externalized this idea and experienced it as an hallucination of their bodily form.

Myers took pains to stress that the recipient of a telepathic idea is by no means a passive contributor to the process in that "perception itself is a form of activity."[77] Moreover, not all of us have, according to him, the power of perception in this form. It is a capacity that only some people can exercise, and those that can, do so involuntarily. Myers described such telepathic perceptions as happening outside of our ordinary, or supraliminal, consciousness and hence outside of our control. He originated the idea that this channel of communication takes place in the subliminal consciousness, a region of mentation that is present but ordinarily inaccessible to us:

> Telepathic and clairvoyant impressions . . . I believe to be habitually received, not by the aid of those sensory adits or operations which the supraliminal [ordinary] self directly commands, but by aid of adits and operations peculiar to the subliminal self, and falling under some system of laws of which supraliminal experience, if standing alone, could give us no information.[78]

Myer's theory of telepathy answered another of the questions that James raised in his review of Schmidkunz's *Psychology of Suggestion*. Why is it, he asked, that "'mind-curers' so often excite the 'imagination' to heal the body, when regular practitioners fail, with their far more impressive incantations and paraphernalia?"[79] The answer, for James, appears to have been because the lay healers, working telepathically, had direct access to the subliminal consciousness of the patient, the same psychical region where many of the pathological ideas that were responsible for their symptoms were located.

The widespread empirical success of the mind-cure movement, Myers's theories about the nature of the subconscious mind, and Janet's clinical case studies appear to have formed, for James, a mutually reinforcing aggregate of facts and insights. He believed that, together, these three different strands of practical and theoretical expertise comprised an important new medical paradigm. Nor were the ramifications of this new paradigm limited to its practical, therapeutic value; it also stimulated a novel avenue of exploration within James's philosophy of knowledge.

A Wild World

James's 1890 essay "The Hidden Self" contained what appears to have been his first public exposition of this tentative new line of thought. The opening paragraphs, unlike the rest of the essay, were not about Janet's work. They dealt instead with more general philosophical reflections about the relationship between scientific and mystical knowledge. Here James explicitly stated that his own opinions about such matters had changed considerably "in the past few years."[80]

In the introductory passages to this essay James argued for the importance of research into the "Unclassified Residuum," or the "dust-cloud of exceptional observations" that float outside of the "accredited and orderly facts of every science." He argued that these phenomena, the ones that conventional science refuses to entertain, are often the source of the most profound insights. He stated that "*anyone* will renovate his science who will steadily look after the irregular phenomena. And when the science is renewed, its new formulas often have more of the voice of the exceptions in them than of what were supposed to be the rules." Of all these exceptions to the current rules, he continued, the mass of phenomena "generally called *mystical*" are the most neglected.[81] However, he insisted, it is just these types of experience that often provide the most fertile seeds for new scientific ideas:

> Repugnant as the mystical style of philosophizing may be (especially when self-complacent), there is no sort of doubt that it goes with a gift for meeting with certain kinds of phenomenal experience. The writer has been forced in the past few years to this admission; and he now believes that he who will pay attention to the facts of the sort dear to mystics, while reflecting upon them in academic-scientific ways, will be in the best possible position to help philosophy.[82]

As the primary illustration of this new conviction of his, James took the "most recent and flagrant example" of "'animal magnetism,' whose facts were stoutly dismissed as a pack of lies by academic medical science the world over, until the non-mystical theory of 'hypnotic suggestion' was found for them."[83] Now, he professed, these facts have been "admitted to be so excessively and dangerously common that special penal laws, forsooth, must be passed to keep all persons unequipped with medical diplomas from taking part in their production."[84]

In the two decades or so since James wrote his review of Liébeault's work, his allegiances had shifted considerably. Far from urging the medical profession to "take possession" of the practices associated with animal magnetism, James was now berating them for carrying out his earlier instructions too enthusiastically. He had come to believe that, when it came to therapeutics, the scientific professionals were not the only authority worth consulting. His youthful self had seen no reason why "educated medical men" could not simply annex the techniques of mental healing for their own scientific, and thus superior, research. By 1890, however, James had arrived at the conclusion that practice and practitioner could not be so easily separated; the mystical lay healers observed phenomena that the "scientific-academic mind" just could not see. In "The Hidden Self" James generalized that "to no one mind is it given to discern the totality of Truth. Something escapes the best of us, not accidentally, but systematically, and because we have a twist. . . . Facts are there only for those who have a mental affinity with them."[85] He posited that every type of observer has an unavoidable blind spot, the same sentiment that was to become the basis of one of his well-known addresses, "On a Certain Blindness in Human Beings," which he first delivered two years later.[86]

James became increasingly convinced, moreover, that these new "mystical" medical facts could not be accommodated within the bounds of conventional scientific thinking. When he first encountered the world of psychical research he saw no reason why such observations would not, one day, be reconciled with the assumption that "the world's body [is] all mechanical."[87] A letter James sent to his friend Robertson, in 1885, illustrates his faith at that time that, ultimately, science would swallow up all of the psychical researchers' anomalous observations and reveal how they are "conditioned by general laws":

> If once these facts show themselves susceptible of serial, & statistical, and other modes of treatment familiar to scientific men, the barrier will fall at a stroke, and we shall hear much less about their rarity and the difficulties of ascertaining them accurately. The very non-teleological character which seems to stick out as an aggregate description of them, will, instead of being a stumbling block as hitherto, become the chief incentive to interest, as showing their analogy with other natural processes, conditioned by general laws.[88]

James argued that the inability of the psychical researchers to reproduce the phenomena they studied at their command would, once the formula for their patterns of occurrence became clear, no longer be a source of frustration but an indication that they were genuine, stubborn facts of nature rather than hoax effects. Over the years that followed, however, James appears to have changed his mind. The actualities pertaining to suggestive therapeutics and faith healing implied, he later asserted, a world that was very different from the "serial, & statistical" rules recognized by medical science; they pointed to a universe where "abnormal personal peculiarities," rather than law-abiding elementary forces, were "the order of the day."[89]

In a lecture on "Exceptional Mental Phenomena" that he composed during the mid-1890s, for example, James observed that the "ordinary medical mind . . . has its cut and dried classifications and routine thera-peutic appliances of a material order"; and such minds find the "dramatic and humoring and humbugging relation of operator to patient" that is characteristic of psychotherapeutic investigations "profoundly distaste-ful." These "orderly characters" of the medical profession, he explained, "don't wish a *wild* world; a world where tomfoolerly seems as if it were among the elemental and primal forces. They are perfectly willing to let such exceptions go unnoticed and unrecorded. They are in too bad form for scientific recognition." James concluded that the phenomena in ques-tion were "so lawless and individualized that it is chaos come again."[90]

A few years later, in 1898, James returned to this same theme in a speech he delivered in defense of the local mind-cure community, whose livelihood was threatened by recently proposed state regulation. James's address, which will be discussed in more detail in chapter 5, opposed the imposition of restrictive licensing legislation, which had been initi-ated by the orthodox physicians. Doctors, James explained to the Mas-sachusetts Committee on Public Health, were too closed-minded. They failed to grasp that they can perceive only one portion of the total truth about healing. There are many different specimens of mind and each type, James insisted, is necessarily "partly perceptive and partly blind. Even the very best type is partly blind. There are methods which it can-not bring itself to use."[91]

Members of the medical profession were blinded, James implied, by the constraints of their medical philosophy. Wedded as they were to ex-clusively "chemical, anatomical [and] physiological information," they could not see the potential of any treatment that seemed to rely on "per-sonality" as one condition of success along with "impressions and intu-itions." The mind-curers were demonstrating that the "therapeutic re-lation may be what we can at present describe only as a relation of one *person* to another *person*," and these "vital mysteries, . . . these personal relations of doctor and patient, . . . these infinitely subtle operations of Nature," claimed James, constituted a valuable new "department of med-ical investigation."[92]

By the late 1890s, when he delivered his address on the proposed med-ical registration bill, James had begun to describe the mind-cure move-ment, with its focus on the patient as an individual person rather than merely a collection of universal, elementary physiological processes, as "a religious or quasi-religious movement."[93] Crucially, he insisted, mind cure was effective from a practical, therapeutic perspective, and those who practiced it enjoyed physical and not just spiritual regenera-tion. When he wrote his lectures on *The Varieties of Religious Experience* at the turn of the century, he assessed, in some detail, the broader, philosophical significance of these observations.

In *The Varieties* James began by describing how the "'scientists' or 'positivists,' as they are fond of calling themselves" explain religious thought as an "atavistic reversion" to a more primitive way of thinking. If asked, James asserted, they would

> probably say that for primitive thought everything is conceived of under the form of personality. The savage thinks that things operate by personal forces, and for the sake of individual ends. For him even external nature obeys individual needs and claims, just as if these were so many elementary powers. . . . [Whereas] science . . . has proved that personality, so far from being an elementary force in nature, is but a passive resultant of the really elementary forces, physical, chemical, physiological, and psycho-physical, which are all impersonal and general in character. Nothing individual accomplishes anything in the universe save in so far as it obeys and exemplifies some universal law.[94]

Up until this point these observations were essentially a reprisal of the comments James made, as a young student, in his review of Huxley's lectures. Thirty or so years later, however, he no longer found himself able to privilege one perspective over the other. He went on to explain how the mind-cure movement had succeeded in subverting science's claims to intellectual superiority.

First, James described how positivistic scientists typically appeal to their "strict use of the method of experimental verification" to assert the supremacy of their approach: "Follow out science's conceptions practically, they will say, the conceptions that ignore personality altogether, and you will always be corroborated." He summarized their claim as follows: "The world is so made that all your expectations will be experientially verified so long, and only so long, as you keep the terms from which you infer them impersonal and universal."[95] James then introduced mind cure with "her diametrically opposite philosophy which sets up an exactly identical claim":

> Live as if I were true, [the religion of mind cure] says, and every day will practically prove you right. That the controlling energies of nature are personal, that your own personal thoughts are forces, that the powers of the universe will respond directly to your individual appeals and needs, are propositions which your whole bodily and mental experience will verify.[96]

James argued that experience does largely verify these "primeval religious ideas," as witnessed by the growth of the mind-cure movement
that does not spread by "proclamation and assertion simply, but by
palpable experiential results." Based on the abundance of regenerative
experiences enjoyed by those who were converted by the mind-cure
doctrines, James affirmed his belief that "the claims of the sectarian scientists are, to say the least, premature." These alternative experiences
"plainly show the universe to be a more many-sided affair than any sect,
even the scientific sect, allows for."[97]

James ended his analysis with a statement that anticipated the essence of his later lectures on the philosophical system of pragmatism. In
his lectures on *Pragmatism* James argued that there are many different
philosophical paths to truth, and that while the tenets of these philosophies may contradict each other, they may each, in their own way, produce results that are useful, or have a "cash value" for humanity: "Any
idea that will carry us prosperously from any one part of our experience
to any other part, linking things satisfactorily, working securely, simpli-
fying, saving labor; is true for just so much, true in so far forth, true *in-
strumentally.*"[98] Similarly, in *The Varieties* he wrote:

> The obvious outcome of our total experience is that the world can
> be handled according to so many systems of ideas, and is so handled
> by different men, and will each time give some characteristic kind
> of profit, for which he cares, to the handler, while at the same time
> some other kind of profit has to be omitted or postponed. Science
> gives to all of us telegraphy, electric lighting, and diagnosis, and suc-
> ceeds in preventing and curing a certain amount of disease. Religion
> in the shape of mind-cure gives to some of us serenity, moral poise,
> and happiness, and prevents certain forms of disease as well as sci-
> ence does or even better in a certain class of persons.[99]

James defended his position with an analogy from the world of math-
ematics:

> And why, after all, may not the world be so complex as to consist of
> many interpenetrating spheres of reality, which we can thus approach
> in alternation by using different conceptions and assuming different
> attitudes, just as mathematicians handle the same numerical and spa-
> tial facts by geometry, . . . by algebra, by the calculus, . . . and each
> time come out right?[100]

This willingness to accord both "science" and "religion" a certain degree
of respect regarding our understanding of the workings of the natural

world represented a significant departure from James's earlier point of view. As a younger man he had conceived of the "unseen world" of religion as a deeper dimension of meaning, one that would provide an explanatory *moral* context for the events that take place in the natural world but not challenge the intellectual authority of the mechanical, impersonal laws of nature. As his essay "The Psychology of Belief" (1889) testified, James was originally concerned with religion as a means of satisfying both the *emotional* needs of the believer and their "impulse to take life strivingly," an impulse that he declared was "indestructible in the race."[101] Similarly, in a book review that he wrote in 1875, James described the decision to place one's trust in the existence of "the world 'behind the veil' of religion" and asserted that "any one *to whom it makes a practical difference* (whether of motive to action or of mental peace) is in duty bound to make it."[102] In this way, James's earliest arguments in defense of religious belief appealed to its role in providing both comfort and a life purpose or stimulus. He did not, however, credit religion with the ability to satisfy our *analytical* need to understand the way the natural world operates. In that province he conceded, as late as 1889, "It is undeniably true that materialistic, or so-called 'scientific,' conceptions of the universe have so far gratified the purely intellectual interests more than the more sentimental [religious] conceptions."[103]

By 1897, however, when he published his essay "What Psychical Research Has Accomplished," James had come to adopt a much more charitable attitude toward the intellectual contributions of the "individual fate-lore of religion." He had become convinced that "the Spirit and principles of science are mere affairs of method; there is nothing in them that need hinder science from dealing successfully with a world in which personal forces are the starting-point of new effects."[104] James concluded that the

> systematic denial on science's part of personality as a condition of events, this rigorous belief that in its own essential and innermost nature our world is a strictly impersonal world, may, conceivably, as the whirligig of time goes round, prove to be the very defect that our descendants will be most surprised at in our own boasted science.[105]

In this essay James credited the field of psychical research with bringing him to this new standpoint. The provocative "*facts of experience*" associated with this field included, for James, the successes of the faith healers and the clinical findings of Janet, whose methods he described as "those of a psychical researcher," as well as the travails of those such as Myers who were members of the Society for Psychical Research.[106]

Only after he became acquainted with Janet's new theories of psycho-pathology, and increasingly convinced by the mind-cure movement, did James come to reject the assumption that religion should defer to a mechanistic science on theoretical matters. The empirical results of the mind-curists legitimated the more general conceit of religious episte-mology as James understood it, namely, the assumption that personal forces effect real change in the world.

These "wonder-mongers" produced outcomes that were of profound importance to James: the specific wonders of therapeutic success. James developed the philosophical insights he presented in his *Pragmatism* lectures over a number of years, and when these ideas about the inter-relations of science and religion surfaced in embryonic form, whether in private correspondence or public writings, it was in the context of medical facts and concerns. It was medical science and medical religion that were at the forefront of his mind. The prevention of disease and the promotion of health were the paradigmatic territory within which James came to rethink and realign his epistemological loyalties.

In this sense, James's hygienic agenda continued unabated during the second half of his career, albeit from a more pluralistic perspective. In-deed, he wrote to a friend in 1902 that he believed "more than ever that *health* is the only good." He added, however, that "all kinds of health, bodily mental and moral are essentially the same, so that one can go at them from any point."[107] A letter to another correspondent, sent a few weeks earlier, indicated the source of his inspiration. There he remarked: "The mind curers have made a great discovery—viz. that health of soul and health of body hang together and that if you get *right*, you get right all over by the same stroke."[108] In his youth James had conceived of sci-ence and religion as pertaining to two separate domains of being; the former governed the mechanical wisdom of the world's body and the lat-ter spoke to its soul of moral meaning. He subsequently came to see sci-ence and religion, body and soul, as interpenetrating pathways through the same wholeness of existence.

CHAPTER 4

Energy and Endurance

Mortal Disease, Morality, and God

Throughout his career, James's publications on ethics and religion were animated by a singularly vital theme. The concept of energy featured consistently in these works and was fundamental to his construction of moral and religious belief systems.[1] This theme appears to have owed much to his close personal identification with the diagnosis of neurasthenia from the late 1870s onward and his conviction that nervous exhaustion was to blame for many of his troubles.[2] He constantly monitored his own levels of nervous energy, ever mindful of the impact of various activities and choices. The usual neurological suspects, such as diet, work habits, and climate, were all subject to his scrutiny, but so too was an entirely different sphere of experience. James's fixation with energy extended beyond the conventional concerns of his medical associates and into the realm of metaphysical inquiry.

Materialism summoned for James the enervating prospect of "a world of fundamentally trivial import," a "simple brute actuality" that was incapable of inspiring any kind of effortful activity. He contrasted this with the faith that there exists something more: a "world behind the veil." James insisted that for those who believe this to be, in its innermost essence, a "radically moral universe," "*energy*" is the watchword.[3] In his address "The Moral Philosopher and the Moral Life" (1891), James claimed that even the "religion of humanity," which had stirred in him a degree of youthful devotion, failed to awaken the truly "strenuous mood." Without a God, he argued, "the appeal to our moral energy falls short of its maximal stimulating power." This ultimate stimulus was essential, he explained, if we are to gracefully endure the agonies of evil, our burdens of pain.

Our attitude towards concrete evils is entirely different in a world
where we believe there are none but finite demanders, from what it
is in one where we joyously face tragedy for an infinite demander's
sake. Every sort of energy and endurance, of courage and capacity
for handling life's evils is set free in those who have religious faith.[4]

This description of God as an "infinite demander" is typical of James's
earlier writings. In his correspondence from the 1860s and 1870s James
explicitly endorsed elements of Stoicism, and his own abstract representations of God as an "infinite mind" were reminiscent of the Stoic notion
of the "*logos,*" an intellectual and moral code that underlies and governs
the entire visible universe.[5] For Stoic philosophers, their ethical aim was
to live in accordance and harmony with this guiding *logos* or universal
"Reason," and James depicted these obligations as "orders" issuing from
a metaphysical "General."[6] His faith was, in his own words, "a cold activity" that did not provide "sympathy" or any "personal communication"
with the "soul of the World."[7] In a private letter from 1882 he described
his god as "an Ideal to attain and make probable" that he was "less and
less able to do without."[8] James proposed, out of personal necessity it
would seem, a taskmaster deity whose expectations of moral compliance
were a spur to action. His was a removed and authoritarian creator who
required much and offered nothing but instruction in return.

James appears to have maintained this sense of a diffuse and distant
divinity for most of his life. In 1886, for example, he wrote to a Christian acquaintance that he did not share her faith, as she understood it,
and instead felt himself "bound in the fetters of individual morality and
the voluntary life, which, for me now seems wholesomer than for me
now a life of surrender to a higher power would be."[9] Similarly, as late
as 1896 he still avowed that he could not call himself a Christian and indeed went further "in not being able to tolerate the notion of a selec-
tive personal relation between God's creatures and God himself as any
thing ultimate."[10]

Six years later however, in his Gifford Lectures on *The Varieties of
Religious Experience* (1902), James explored and embraced a very different kind of faith and a very different kind of God. This was a God
from whom issued genuine help and not merely commands. Energy
was still at the center of this alternative theology, but its derivation and
ultimate origins were significantly reconfigured. The link James made
between religion, energy, and the evils of pain and illness remained constant though, and it appears to have been critical to his developing ideas.
James's later allegiances to a more intimate and active God were forged

in the crucible of his own experiences of the most profound "concrete evils," namely, those of mortal disease and the fear of his own impending death.

The final few years of the nineteenth century were a time of serious personal health problems for James.[11] In June of 1899 he severely strained his heart, for the second time, while hiking in the Adirondacks, and this was the beginning of an extended period during which he was submitted to the "vile, inert, cowardly professional-invalid life."[12] He and Alice traveled to Europe so that James could undergo treatment at the baths in Nauheim, Germany, which were a renowned cure for such heart problems. While there he wrote to his mother-in-law describing his ambivalent attitude toward the "'Kur' life":

> Here I am, a part of this great organization of cormorants & leeches, all drawing their sustenance from the victims of disease, but apparently making good return in the way of physical regeneration. Nevertheless the spectacle has something morally sickening and repulsive in it, and the sooner one gets out the better.[13]

James was quietly optimistic that the treatment would improve his condition but equally disgusted by the listlessness of a life with "nothing to do all day long except drawl and dawdle, bathe and eat" where one is "forbidden to make any rapid movement, to ascend anything, to get excited about anything." By October, however, he was able to report to a family friend that "my heart is said by the doctor to be organically greatly improved. The subjective symptoms of disturbance are still strong, but I live in hope that they may be only 'nervous,' and not worthy of much practical regard. I shall be able, however, to walk but little this winter."[14]

Unfortunately, James's improved state of health did not last long, and the next month he wrote to another correspondent, blaming himself for the deterioration in his symptoms: "I seem to have overdone the activity . . . for little by little I found myself able to walk less, and 2 ½ weeks ago I had a regular thoracic collapse with the formidable chest pain which I had before I left America." He underwent a further series of medicinal baths, this time in London, but although his physician found the condition of his heart to be much better, James confessed that "my discomfort is as great as ever and I can make no exertion of any sort without symptoms of severe distress." Even reading the newspaper or dictating a note, he told another confidant, brought on "acute thoracic distress." His English doctor sent him to recuperate in the "tonic air of West Malvern," but there he had, what he described as, a "bad break-

the thigh the tips of the operator's fingers gradually
glide round to the ulnar aspect of the wrist, so as to
resist the downward and backward movement of the
arms.   This is the operator's *pons asinorum*, but it
should be mastered.

No. 6.—The trunk is flexed forward, without the
knees being bent, and then brought back to the erect

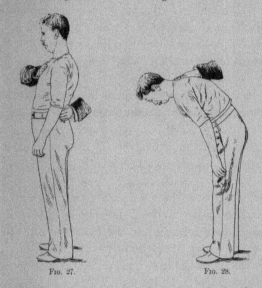

FIG. 27.   FIG. 28.

position.   The operator stands at the patient's side
with one hand over the upper third of the sternum,
and the other supporting the mid-lumbar region
(Fig. 27).   The reverse movement is resisted by
placing one hand over the junction of the cervical
and dorsal portions of the spine (Fig. 28).

4.1 James's treatment at the Nauheim baths was overseen by the respected heart specialist Theodor Schott and included a specific series of regular exercises, one of which
is illustrated above. (W. Bezly Thorne, *The Schott Methods of the Treatment of Chronic
Diseases of the Heart with an Account of the Nauheim Baths, and of the Therapeutic Exercises* [London: J. & A. Churchill, 1896], 53 [Wellcome Collection].) James wrote to his
doctor friend James Jackson Putnam describing his experiences of Schott's methods and
enclosing a copy of the book from which the images are taken: "The gymnastics—I send
you Besley [*sic*] Thorne's book which explains them—are a singularly grateful tonic. It
is astonishing that so few & such slightly energetic movements should give one such a
corporeal feeling of *wohlsein* [well-being] afterwards. I never should have believed it."
(William James to James Jackson Putnam, September 19, 1899, Skrupskelis and Berkeley,
eds., *Correspondence*, 9:49.)

down." The nature of the breakdown appears to have been as much psychological as heart related. James had become convinced that he wasn't going to live much longer.[15]

By December of 1899 James had come to believe that he was fatally ill and to wonder whether he would live to honor his prestigious appointment by the Gifford committee in Scotland to complete a series of lectures on natural religion.[16] He wrote to a family friend that

> in a general manner I can see my way to a perfectly bully pair of volumes, the first an Objective study of "the Varieties of Religious Experience," the second, my own last will and testament, setting forth the philosophy best adapted to normal Religious needs. I hope I may be spared to get the thing down on paper.[17]

A week later he wrote to a confidant, Théodore Flournoy, in an even more pessimistic vein: "I am passing off the scene, have resigned my Gifford Lectureship, and shall certainly be unable to keep my Professorship at Harvard, and shall very likely die with my great Philosophy of Religion buried inside me and never seeing the light."[18] An entry in a notebook that James kept during this period of his life further illustrates his state of mind:

> I find myself in a cold, pinched, joyless quaking state. . . . When I think of the probability of dying soon with all my music in me. My eyes are dry and hollow, my facial muscles won't contract, my throat quivers, my throat heart flutters, my breast and body feel as if they were stale and caked as if stale and caked. . . . My mind is pent in and pinned down to the narrow continual contemplation of my own an annihilation which fills me with a kind of physical dread, . . . tThe increasing pain & misery of more fully developed disease—the disgust, the final strangulation etc., be[g]in to haunt me, I fear them; and they preoccupy my attention the more I think fear them, the more I think about them.[19]

Eventually, however, James's correspondence documented his gradual recovery. His mood appeared to lighten, and he began to refer in the past tense to the "bad breakdown" that he had recently endured.[20] But although James resumed work on his lectures and believed that he had left his most desperate days behind him, their indelible mark remained. To a greater extent than any of the other works that James produced, either before or afterward, *The Varieties of Religious Experience* exemplified the many different ways that philosophy in general, and religious philosophy in particular, was entangled with the concepts of disease and health in James's thought.[21] The subtitle of the published series should perhaps have read "What Religion Can Do for the Sick and Infirm," because this is the sentiment that lies at the heart of the text. The opening chapters, for example, may be read as an exploration of the question that had become of pressing personal importance to James in the depths of his suffering: How best, ethically speaking, should one meet the evil of disease and its ultimate manifestation of death?

In February of 1900, in a letter to his friend Thomas Davidson, who was undergoing treatment for a serious prostate problem, James outlined the options as he saw them: "One can meet mortal (or would-be mortal) disease, either by gentlemanly levity, by high-minded stoicism, or by religious enthusiasm. I advise you, old T.D., to follow my example and try a playful durcheinander [muddle] of all three, taking each in turn *pro rê nata* [according to circumstances]."[22] By the time he completed *The Varieties* James had made his choice, theoretically at least. His manuscript notes concluded, as did his published lectures, that all things considered the more mystical form of religion offers the fullest solution to life's most difficult moments. In the public version, the subtext that such moments comprise the daily torment of the sick and dying is present but muted. In his confidential musings it is pervasive and unmistakable.

In his notebook James dealt first with the merits of Stoicism, the philosophical life path that had held a great attraction for him ever since he first read the *Meditations* of Marcus Aurelius as a young man. He stated that

> the higher religious attitude for a man with fatal disease (and in truth then for every man and at all times) is that of attention voluntarily directed away from his own future, whether in this world or the next; . . . of immersion in whatever objective interests still remain possible to him, . . . silence over his own miseries, contemplation of ~~whatever the better~~ those higher general aspects of existence ~~which phil~~ which his philosophy may furnish him withal, and cultivation of whatever tender ~~optimistic~~ virtues, such as resignation, patience, fortitude, his ethical system ~~offers~~ proposes to him as ideals . . . Practically . . . I have described the Stoic.[23]

This sort of Stoic "remedy" is available in principle, he continued, to anybody "no matter how sick or helpless he may be. You have but to disdain the ~~trembling~~ pinched and mumping sick room attitude, and take your stand on the ~~other~~ out of door one and there you are."[24] But, he conceded, this path is effectively one of self-administered help, and therein lies its limitation. The higher moral order invoked by Stoicism does not offer itself as anything other than a motive: an abstract ideal that acts as a "muscular excitement," or stimulus, for those who believe in its existence. This appeal to self-reliance is ultimately, James acknowledged, doomed to failure:

> Stoicism, with its muscles ~~never relaxed~~ always tense, always ~~tensely sustaining~~ holding its breath in an attitude which is ready to ~~lapse into its opposite, &~~ break down, and at the last extremity always does ~~so lapse~~ break down has in this instability an element of weakness which religion in the more extreme sense has always felt.... [In the] ... radical manifestations [of] religion ... Will is ~~swallowed up in~~ drowned in ~~excitement~~ peace attained.... The ~~muscular~~ hour of muscular tension is over, that of happy relaxation, of calm deep breathing, of a present with no ~~different~~ possibly different future to be on one's guard against has arrived.[25]

The difference between Stoicism and these extreme kinds of religious experience, such as "mystic states" and "methodistical conversion," is, James argued, that the former merely ignores life's "negative element" whereas the latter expunge it. To achieve this, he noted, "it would seem that directly vital relations with a supersensual world were required"; in order to acquire this passionate peace in the midst of "the sensible pressure of the world's ill and danger" the "inflow of super-personal help" is necessary. This religious way is "the more complete, undoubtedly, and the latter persons are the more evolved," James concluded. He added that his "interpretation makes for pluralistic Zielstrebigkeit [purposefulness], since the goal of all alike is equanimity in the presence of the world's ill."[26] Essentially, James made what had undoubtedly always been an important life challenge for him into a universal goal for all humankind, namely, the ability to maintain a calm and enduring fortitude in a world that contains the evils of pain and illness. This task necessitated, he believed, different metaphysical loyalties for different types of people, hence the diversity of religious experiences included in his study.

In the final version of the lectures, the personal context in which they were conceived, the confines of the sick room, is not immediately discernible. In his second lecture, entitled "Circumscription of the Topic," James outlined the scope of his project. He first made it clear that he was interested only in personal, as opposed to institutional or ecclesiastical, manifestations of religion. He further narrowed the field of inquiry to the more instructive forms of religiosity. As far as James was concerned this meant the "more energetically religious moods." Finally, he described to his audience how "morality pure and simple" may be differentiated from "religion" and thus eliminated from the detailed analysis that followed. It is at this juncture that the autobiographical slant of the lectures became visible for the first time. James, in his delineation of morality and religion, invoked the same argument detailed above in his notebook. Fundamentally, he argued, the difference between moral and religious faiths is their accessibility and value to the sick and dying. He explained that

> the athletic attitude [of the moralist] tends ever to break down, and it inevitably does break down even in the most stalwart when the organism begins to decay, or when morbid fears invade the mind. To suggest personal will and effort to one all sicklied o'er with the sense of irremediable impotence is to suggest the most impossible of things. . . . And here religion comes to our rescue and takes our fate into her hands. There is a state of mind, known to religious men, but to no others, in which the will to assert ourselves . . . has been displaced by a willingness to . . . be as nothing in the floods and waterspouts of God. In this state of mind, . . . the time for tension in our soul is over. . . . Fear is not held in abeyance as it is by mere morality, it is positively expunged and washed away. . . . [Religion] ought to mean nothing short of this new reach of freedom for us.[27]

James explicitly rooted his definition of religion in its ability to assist us with the earthly trials of pain and suffering. He stated that "religious feeling is thus an absolute addition to the Subject's range of life. It gives him a new sphere of power." In the introductory section of the lectures the "new sphere of power" is described, primarily, as a freedom from morbid fear and dread. It takes the form of happiness, a kind of happiness that he considered unique to religion by virtue of its completeness. Religious happiness was not merely a form of "escape"; instead, it comprised a total surrender to the evils of life. According to James, while outwardly consenting to these evils, as "a form of sacrifice," the truly religious person inwardly knows that this evil has been "permanently overcome." How religion achieves this seeming magic act, how it "falls on the thorns and faces death, and in the very act annuls annihilation," he declared to be "religion's secret."[28]

The Divided Self

The seeming paradox of religious submission and the new lease on life that it brings in its wake were the focus of the middle chapters of *The Va-*

rieties. There James constructed a psychological explanation of why religious conversion is frequently accompanied by an increase in vitality, which he maintained was its distinguishing trait. He began by developing his idea that different types of people require different types of faith, and he classified these religious needs into two alternative groups.

The "healthy-minded" were those who have a "constitutional incapacity for prolonged suffering" and are thus able to sustain a sunny, optimistic type of belief system. According to James, these sorts of philosophy either disregard the existence of evil altogether or consider it "unmanly and diseased" to dwell on the pain that shadows all human life. There are others, however, whom James described as "sick souls," meaning "those persons who cannot so swiftly throw off the burden of the consciousness of evil, but are congenitally fated to suffer from its presence."[29]

He invoked the recent psychological concept of a "threshold" as a designation for the point at which one state of mind passes into another. In this vein, he suggested, "we might speak of a 'pain-threshold,' a 'fear-threshold,' a 'misery-threshold,' and find it quickly overpassed by the consciousness of some individuals but lying too high in others to be often reached." Whereas the "sanguine and healthy-minded live habitually on the sunny side of their misery-line, the depressed and melancholy live beyond it, in darkness and apprehension."[30] It is impossible for those afflicted in this way, James explained, implying that he was one of them, to ignore the metaphysical significance of evil:

> The fact that we *can* die, that we *can* be ill at all, is what perplexes us; the fact that we now for a moment live and are well is irrelevant to that perplexity. We need a life not correlated with death, a health not liable to illness, a kind of good that will not perish, a good in fact that flies beyond the Goods of nature.[31]

In the passages that followed James shared ever more intense biographical reports of the sick soul. He averred that, for some, their pessimism is so great that they are forced in spite of themselves to ignore all that is good. Worse still than this "incapacity for joyous feeling," they may suffer an "active anguish" that takes various forms: "sometimes more the quality of loathing; sometimes that of irritation and exasperation; or again of self-mistrust and self-despair; or of suspicion, anxiety, trepidation, fear." For these melancholics, the world looks "remote, strange, sinister, uncanny. Its color is gone, its breath is cold."[32]

James asserted that when disillusionment had progressed this far, no return to ordinary happiness, with its "simple ignorance of ill" was possible. Either happiness fails to reoccur at all in an "acute form," or it reappears as "something vastly more complex, including natural evil as one of its elements, but finding natural evil no such stumbling-block and terror because it now sees it swallowed up in supernatural good." This process, James continued, is "one of redemption, not of mere reversion to natural health" and is experienced as a "second birth." He concluded that religious systems such as Buddhism and Christianity, that actively address the evils of "sorrow, pain, and death" in this way, are the "completest." He described them as "religions of deliverance" in that "man must die to an unreal life before he can be born into the real life"[33]

In his subsequent chapter on "The Divided Self" James explained, in technical, psychopathological terms, the origins of the mental torment that defines the "sick souls" who are in need of these "twice-born" religions. He opened with a discussion of what he described as "heterogeneous personality." Although some people are born with an "inner constitution" that is "harmonious and well balanced from the outset," others, he explained, "are oppositely constituted; and are so in degrees which may vary from . . . a merely odd or whimsical inconsistency, to a discordancy of which the consequences may be inconvenient in the extreme." These more serious manifestations of heterogeneity were exemplified, for James, in those whose "spirit wars with their flesh," those who "wish for incompatibles," and those in whom "wayward impulses interrupt their most deliberate plans." In essence "their lives are one long drama of repentance and of effort to repair misdemeanours and mistakes."[34]

James appears to have borrowed the term "heterogeneous personality" from the work of an American neurologist, Smith Baker (1850–1922), whom he cited in this section of *The Varieties*. In 1893 Baker published a paper where he argued for the etiological significance of this type of psychological splitting. He believed that it could account for a wide range of mental and neurological disease, owing to the considerable "wastage of personal force" associated with the internal battles occasioned by such "intra-personal belligerency." He asserted that "alienists and neurologists find evidence of some such painful contest probably in nine tenths of those who need their services." Such battles were particularly common, he stated, "in connection with a consideration of the sexual nature and its derangements . . . [and the] struggle to make dominant the better aspects of personality." He also noted that "ecclesiastical biography" affords multitudinous examples of those struggling with a heterogeneous personality.[35] Baker surmised that the condition was the result of inheritance, where the traits of character of "incompatible and antagonistic ancestors" were preserved alongside each other.[36] James, however, refused to be drawn one way or the other on this particular aspect of Baker's theory. Instead he explained that, regardless of how such internal divisions may come to be constituted,

> in all of us . . . to a degree the greater in proportion as we are intense and sensitive and subject to diversified temptations . . . does the normal evolution of character chiefly consist in the straightening out and unifying of the inner self. The higher and the lower feelings, the useful and the erring impulses begin by being a comparative chaos within us [and] must end by forming a stable system of functions in the right subordination.[37]

"To find religion" was, according to James, one route to this unified state. He was careful to emphasize, however, that conversion is not the only example of what is a "general psychological process," which may take place with "any sort of mental material" and does not necessarily assume a religious form. He gave other examples of "the new birth" that may move the individual in the opposite direction, from religion into incredulity, or from "moral scrupulosity into freedom and license," or be driven by an entirely different passion, such as love, ambition, revenge, or patriotic devotion.[38] James was clear that he was advancing a universal model of psychopathology. In essence, this model was an extension of the hypothesis that certain forms of mental inhibition are detrimental to health, which featured in his *Talks to Teachers*. In both lecture series he deployed a psychoneurological framework that stressed the harmful consequences of clashing internal ideas and impulses.

In his chapter in *The Varieties* entitled "Conversion," James expanded upon his conception of the component parts that comprised a divided self. He drew on the psychological language of "association" to explain that

> a man's ideas, aims, and objects form diverse internal groups and systems, relatively independent of one another. Each "aim" which he follows, awakens a certain specific kind of interested excitement, and gathers a certain group of ideas together in subordination to it as its associates; and if the aims and excitements are distinct in kind, their groups of ideas may have little in common.[39]

James clarified that a person may possess several distinct aims each associated with its own collection of ideas, and they may, accordingly, be occupied at one moment with one set and at other times with a different set. Ordinarily, he continued, when we pass from one of our aims to another these alterations of character are not remarked upon because "each of them is so rapidly succeeded by another in the reverse direction." In cases where one aim becomes dominant and stable, however, James asserted that we tend to refer to the phenomenon, "and perhaps to wonder at it, as a 'transformation.'"[40] He characterized these alterations between different sets of aims and ideas, whether temporary or permanent, as shiftings of the focus of internal energy or excitement:

> Now there may be great oscillation in the emotional interest, and the hot places may shift before one almost as rapidly as the sparks that run through burnt-up paper.... Or the focus of excitement and heat, the point of view from which the aim is taken, may come to lie permanently within a certain system; and then, if the change be a religious one, we call it a *conversion*, especially if it be by crisis or sudden.[41]

In the absence of any such conversion or transformation, James stated that the psychical division may be complete, when a person's primary aims alternate and take turns, or less complete when two or more different aims coexist, but there is only one that "practically holds the right of way and instigates activity, whilst the others are only pious wishes, and never practically come to anything."[42] Crucially, throughout his detailed exposition of the divided self, James described the division in terms of different aims, ambitions, desires, and volitions. In short, at its core, the "divided self" was "a divided will," a phrase that James used explicitly in places.[43]

It is instructive to compare and contrast James's notion of the divided self with the theories of his European contemporaries such as Alfred Binet and Pierre Janet. For these psychologists the self was fractured along the fault lines of memory. In their cases of multiple personality each of the individual personalities was held together by its own set of memories; every secondary personality had a separate train of remembrance that defined and demarcated it from the others. The possibility that these alternative selves might possess knowledge or aims that were valid or superior to those maintained by the primary consciousness was not given any serious thought. They were peculiar anomalies to be toyed with, or cured (eradicated), not listened to or taken seriously in their own right. In the notable case of Marie that was discussed in chapter 3, for example, Janet's therapeutic approach involved modifying the memories of her secondary personality, rather than contesting its volitions or facilitating their accommodation within the wider self of the patient. These French theorists, and others such as the Swiss alienist Eugen Bleuler, advanced psychopathologies of memory in which the concept of "the will" was effectively redundant.[44]

In contrast, memory played a much less important role within James's divided self. The two halves were partitioned by their intentions, not their recollections. The relative accessibility of different sections of mental content, or memory, appears to have been a supplementary rather than defining characteristic within his schema. In the examples James cited of heterogeneous personality in *The Varieties*, the subject was aware, at all times, of their dual nature. It was only the thought processes via which one side gained ascendency over the other that were sometimes purported to have happened, behind the scenes, in a deeper region of consciousness. When he discussed the psychological activity that governs the "shifting of men's centres of personal energy," he explained that these phenomena are "partly due to explicitly conscious processes of thought and will, but as due largely to the subconscious incubation and maturing of motives deposited by the experiences of life. When ripe, the results hatch out, or burst into flower."[45] Within James's formulation there would be nothing to prevent an entire cluster of aims and ideas, one side of a divided will, developing out of sight within the subliminal consciousness, but this was not a prerequisite. His conceptualization appears to have been inspired by the ideas of the neurologist Benedikt together with those of Baker, who both made an internal conflict between two wills central to their model of mental disease.[46]

James's emphasis on the divided self as a divided will was fundamental to his understanding of the act of conversion, or cure by religion. As described above, he believed that the resolution of a divided self may happen gradually, "consisting of the building up, piece by piece, of a new set of moral and spiritual habits," or suddenly as a result of subconscious mentation.[47] Following the lead of fellow psychologist and former pupil Edwin Starbuck (1866–1947) in his book *The Psychology of Religion* (1899), James distinguished between these two types of conversion, naming them "the *volitional type* and the *type by self-surrender*."[48] James also agreed with Starbuck, however, that even the vast majority of volitional conversions involve, ultimately, an act of surrender, in that the personal will must be given up: "In many cases relief persistently refuses to come until the person ceases to resist, or to make an effort in the direction he desires to go."[49]

James explained that the reason this surrender appears to be necessary is because otherwise the individual's conscious striving can get in the way of the process of psychological resolution, or conversion, that is happening in the subconscious region. In his words: "When the new centre of personal energy has been subconsciously incubated so long as to be just ready to open into flower, 'hands off' is the only word for *us*, it must burst forth unaided!"[50] For this reason, he argued, self-surrender has been and always must be regarded as the vital turning-point of the religious life. . . . One may say that the whole development of Christianity in inwardness has consisted in little more than the greater and greater emphasis attached to this crisis of self-surrender. . . . Psychology and religion are thus in perfect harmony up to this point, since both admit that there are forces seemingly outside of the conscious individual that bring redemption to their life.[51]

James turned the religious act of throwing oneself on the mercy of God into a psychotherapeutic procedure.[52] This act of self-surrender facilitated, in psychological terms, an inner state that was conducive to the termination of the internal battle between the two idea systems. A nervous system that was no longer crippled by the opposing volitions of two competing centers of psychological activity could then operate more freely; the flow of nervous energy was no longer inhibited, and the subject enjoyed a newfound sense of power and "regenerative phenomena."[53]

In Starbuck's text this process was described as a natural part of adolescence. James agreed that his conclusion seemed sound and that conversion was, "in its essence a normal adolescent phenomenon, incidental to the passage from the child's small universe to the wider intellectual and spiritual life of maturity." Their studies differed considerably, however, as to the emphasis they placed on the various consequences of the experience. Starbuck mentioned, in passing, that young converts experience a rush of nervous energy, but for him the true significance of the event lay in the newly mature outlook that adolescents gain. For the first time they take their place as adults in society, which meant, for Starbuck, that they cease to focus exclusively on themselves and their own needs and begin to display altruism and to think about how they might contribute to the wider community. James, meanwhile, appeared, in this context, to be much less interested in this change in values and much more interested in the immediate energy-related consequences of the transition to religiosity.

The other point on which James's study parted company from Starbuck's analysis was the question of whether any conversion could be considered the work of a higher spiritual agency. Starbuck suggested that the conversion experience should be understood in the physiological context of the rapid development of the nervous system of the adolescent:

Among the changes in adolescence those in the anatomy of the nervous system furnish an important background for understanding the spiritual phenomena. . . . It is instructive to note the correspondence in time between the rapid psychic development at adolescence and the increment in the growth of a new band of association fibres. It seems that the great psychic awakening may be conditioned by either of two things: *either there is a new crop of nerve branches which rapidly reach functional maturity, or those which have already matured come suddenly into activity.* This furnishes the basis of the power of the mind . . . for intellectual grasp, and for spiritual insight.[54]

From Starbuck's perspective the new powers that develop at the time of conversion are the by-products of newly operational nerve networks. James did not comment directly on his neurological account. He did, however, distance his investigations from the typical adolescent conversions "of very commonplace persons" such as "Dr. Starbuck here has in mind." James professed himself more interested in the "sporadic adult cases." Of such cases, those where religious feelings ripen late in life suggested to him "more than any others the idea that sudden conversion is by miracle."[55] As his lectures drew to a close, James outlined an explanation of such experiences that could accommodate both natural and supernatural accounts.

Superhuman Life

It is evident from James's published writings and correspondence that he had been developing his ideas about the psychology of conversion for several years prior to his presentation of the Gifford Lectures. In one letter, from 1897, he clarified his position on the role of the supernatural:

> I am quite willing to believe that a new truth may be supernaturally revealed to a subject when he really *asks*. But I am sure that in many cases of conversion it is less a new truth than a new power gained over life by a truth always known. It is a case of the conflict of two *self-systems* in a personality up to that time heterogeneously divided, but in which, after the conversion-crisis, the higher loves and powers come definitively to gain the upper hand and expel the forces which up to that time had kept them down in the position of mere grumblers and protesters & agents of remorse and discontent.[56]

This undogmatic perspective on Godly intervention was a notable feature of James's *Varieties of Religious Experience*. There he relied on the concept of a subconscious realm of mentation to mediate between science and theology, stating that "the theologian's contention that the

religious man is moved by an external power is vindicated, for one of
the peculiarities of invasions from the subconscious region is that they
take on objective appearances, and suggest to the Subject an external
control." In other words, when a subject feels as if they have communed
directly with a higher power, it may actually be the "higher faculties of
[their] own hidden mind" that have made their presence felt.[57] Equally,
however, James insisted that it is "logically conceivable that *if there be*
higher spiritual agencies that can directly touch us, the psychological
condition of their doing so *might be* our possession of a subconscious
region which alone should yield access to them."[58]

James based this hypothesis on a specific model of the subconscious,
which he acknowledged in the lectures as belonging to the psychical
researcher Frederic Myers. Myers proposed that the subliminal region
of consciousness was open-ended and extended, as James put it, into a
wider "cosmic environment" that had the "the character of a 'spiritual
world.'"[59] Myers also attributed another distinctive feature to the subconscious, and when James reviewed his, posthumously published, *Human Personality and Its Survival after Bodily Death*, this was one of the
elements of Myers's text that he singled out for discussion:

> From its intercourse with this spiritual world the subliminal self of
> each of us may draw strength, and communicate it to the supralimi-
> nal [ordinary, waking] life. The "energizing of life" seems, in fact, to
> be one of its functions. . . . The regenerative influences of prayer and
> of religious self-surrender, the strength of belief which mystical ex-
> periences give, are all ascribed by Myers to the "dynamogeny" of the
> spiritual world, upon which we are enabled to make drafts of power
> by virtue of our connection with our subliminal.[60]

James borrowed the term "dynamogeny" from physiology and specified, elsewhere, that "the physiologists call a stimulus 'dynamogenic'
when it increases the muscular contractions of men to whom it is applied; but appeals can be dynamogenic morally as well as muscularly."[61]
James, then, understood Myers's subconscious domain as one of physical and moral invigoration, and this interpretation is consistent with
how he himself described the impact of religious encounters throughout *The Varieties*:

> What is attained is often an altogether new level of spiritual vitality,
> . . . and new energies and endurances are shown.[62] . . . What the more
> characteristically divine facts are, apart from the actual inflow of en-
> ergy in the faith-state and the prayer-state, I know not. . . . Our pres-

ent consciousness . . . [and other] worlds of consciousness become continuous at certain points, and higher energies filter in.[63]

For James, the hallmark of religious experiences was the influx of energy that they brought with them.[64] These energies emanating from the subliminal consciousness were described by Myers as "recuperative," and he explicitly linked them with the "*vis medicatrix Naturae*" or healing power of nature.[65] James, similarly, laid considerable stress on the medical importance of religion. In his concluding lecture he returned to the question of what religious experiences bring to life, the "peculiar and characteristic sort of performance . . . that element or quality in them which we can meet nowhere else."[66] In the closing passages of his text, he characterized religious feelings or faith states in the following way:

> The resultant outcome of them is . . . an excitement of the cheerful, expansive, "dynamogenic" order which, like any tonic, freshens our vital powers. . . . We have seen how this emotion overcomes temperamental melancholy and imparts endurance to the Subject. . . . It is a biological as well as a psychological condition. . . . We are obliged, on account of their extraordinary influence upon action and endurance, to class [religions] amongst the most important biological functions of mankind.[67]

Put simply, James contended that religion is something that makes an individual's life healthier, the ultimate moral medication. In one sense there was nothing novel about this portrayal. James had long associated religious faith with energy and endurance and praised its worth to those struggling with the evils of pain and illness. Previously, however, the energies set loose belonged to the individual in question and the role of God was solely to command and inspire. In the final pages of *The Varieties*, James baldly stated his new belief that God plays a more active role and operates as an external physical force, an "actual inflow of energy" as opposed to merely a mental essence. Or as he expressed it, religion is manifest in "body as well as soul":

> Religion, in her fullest exercise of function, is not a mere illuminator of facts already elsewhere given. . . . But it is something more, namely, a postulator of new *facts* as well. The world interpreted religiously is not the materialistic world over again, with an altered expression; it must have, over and above the altered expression, *a natural constitution* different at some point from that which a materialistic world would have.[68]

James's conception of God as a "new force" appears to have been derived, in part, from his decades-long wranglings with evil. He no longer
found himself able to endorse a God who gives meaning and direction
to the physical world but does not intermittently reach in to change it.
This kind of faith, he argued, fails to acknowledge the wisdom of the pessimistic perspective: it "takes the facts of physical science at their face-
value, and leaves the laws of life just as naturalism finds them, with no
hope of remedy, in case their fruits are bad."[69] Earlier, in his section on
sick souls and their melancholy pain, James had distinguished between
two different kinds of evil. The first type, he implied, was amenable to
conventional medical guidance, the physiological laws of health that
preached that disease may be eliminated via careful attention to environmental influences and personal habits; but the second, James alleged,
was not so easily dealt with:

> There are people for whom evil means only a mal-adjustment with
> *things*, a wrong correspondence of one's life with the environment.
> Such evil as this is curable, in principle at least, upon the natural
> plane, for merely by modifying either the self or the things, or both
> at once, the two terms may be made to fit, and all go merry as a mar-
> riage bell again. But there are others for whom evil is no mere rela-
> tion of the subject to particular outer things, but something more
> radical and general, wrongness or vice in his essential nature, which
> no alteration of the environment, or any superficial rearrangement of
> the inner self, can cure, and which requires a supernatural remedy.[70]

James, by this point in his life, had become convinced that such supernatural remedies did exist, for some people at least. In the postscript
to *The Varieties*, he stated his allegiance to what he described as a "super-
naturalism" of the "piecemeal or crasser type." Although not the God of
"popular Christianity," he proposed that there was a God who "actually
exerts an influence, raises our centre of personal energy, and produces
regenerative effects unattainable in other ways." This was an interper-
sonal God of "divine aid coming in response to prayer," who is able to
"get down upon the flat level of experience and interpolate itself piecemeal between distinct portions of nature."[71]

James continued to espouse these beliefs for the remaining years of his
life, and they comprised one aspect of the religious philosophy that he
developed in his Hibbert Lectures, entitled *A Pluralistic Universe*, which
he delivered in 1908. There he discussed the "deeper reaches" of religious
experience, which are familiar to "evangelical Christianity and to what
is nowadays becoming known as 'mind-cure' religion or 'new-thought'":

The phenomenon is that of new ranges of life succeeding on our most despairing moments. [These] possibilities that take our breath away, of another kind of happiness and power, based on giving up our own will and letting something higher work for us, . . . seem to show a world wider than either physics or philistine ethics can imagine.[72]

For James these faith healing testimonies formed an important part of the supporting evidence for his metaphysical beliefs. They helped establish "a decidedly *formidable* probability in favor of the existence of "some form of superhuman life with which we may, unknown to ourselves, be co-conscious." The fact that someone could receive more energy, a new lease on life, precisely at the moment when they've given up, and their "lower being has gone to pieces in the wreck," supported the idea that there is an "invisible spiritual environment from which help comes."[73]

In *The Varieties*, James clarified that such healing energies were not available to everyone, however, and there is no indication that he ever enjoyed such help himself. In his section on conversion, he discussed those who "may be excellent persons, servants of God in practical ways, but they are not children of his kingdom." In these subjects of "'barrenness' and 'dryness'" the religious faculties are "checked in their natural tendency to expand, by beliefs about the world that are inhibitive." In a tone of confession, he included under these beliefs "the agnostic vetoes upon faith as something weak and shameful, under which so many of us to-day lie cowering, afraid to use our instincts." Within these people "their personal energy never gets to its religious centre, and the latter remains inactive in perpetuity." However, he insisted, it sometimes happens that even late in life, "some thaw, some release may take place, some bolt be shot back in the barrenist breast, and the man's hard heart may soften and break into religious feeling."[74]

There is evidence that James continued hoping, to the end of his life, that he might find a way to release such inhibitions within himself. In 1909 he engaged the services of a Christian Science healer, called Mr. Strang, anticipating that he might be able to relieve him of his chest pains, respiratory problems, and nervous prostration, but after twenty-one sessions James ruefully conceded defeat. In a private letter about the course of treatment, written eight months before he died, he concluded that there must be

a certain impediment in the minds of people brought up as I have been, which keeps the bolt from flying back, and letting the door of the more absolutely-grounded life open. They can't back out of their system of finite prudences and intellectual scruples, even though in

words they may admit that there are other ways of living, and more
successful ones.[75]

James, it would seem, was never to enjoy the medicinal benefits of contact with the "superhuman" realm of life. This interest in the healing
powers of religious experience was, however, only one part of a more
comprehensive study of personal energy and its implications for health.
In 1906 James brought these researches together for his presidential address to the American Philosophical Association on "The Energies of
Men."

The Energies of Men

In this address James made his medically oriented interests explicit from
the very beginning. The opening paragraphs constituted a glowing endorsement of those who take "the physician's attitude" in psychology
and a correspondingly withering dismissal of "psychologists who are not
doctors," whose technical terms and descriptions of the human mind
and its "aberrations from normality" are "lame" and inadequate. The
"clinical" or "functional" approach in psychology produced, he stated,
ideas of "far more urgent practical importance" and was "assuredly the
thing most worthy of general study to-day."[76]

He explained that what followed would be a discussion of one particular notion of "functional psychology": "the conception of the *amount
of energy available* for running one's mental and moral operations by."
He was careful to distinguish this psychological energy from "the energies of the nervous system." Although it is undoubtedly connected,
he continued, "it presents fluctuations that cannot easily be translated
into neural terms." He maintained that to "have its level raised is the
most important thing that can happen to a man, yet in all my reading
I know of no single page or paragraph of a scientific psychology book in
which it receives mention."[77] This sense of the limitations of "scientific
psychology" and neurology was central to James's "Energies of Men"
text. Although he still saw these disciplinary contributions as valid, he
was keen to promote, alongside them, the recent insights of one particular medical psychologist, the French investigator Janet whose earlier
work on psychotherapeutic methods had previously inspired James's
own thinking.

In 1903 Janet had begun publishing on an entirely new area of research: his innovative diagnosis of "psychasthenia."[78] In a book that is
still regarded as an impressive volume of clinical observation today, he
laid out his detailed descriptions and hypotheses about a class of pa-

tients whose specific symptoms varied enormously but who shared a common underlying issue: a lowering of their psychological "tension" or, as James put it, "oscillations of the level of mental energy."[79] In contrast to the preexisting laboratory categories of "thresholds" and "impulses and inhibitions" and "fatigue" that operated at the level of nerve fibers and cells, Janet's work on the concept of energy was situated in biographical analysis and focused on the emotions and life experiences of those he examined and treated. (James's investment in his colleague's ideas was evidently personal as much as professional, and he wrote to more than one correspondent describing himself as psychasthenic as opposed to neurasthenic as he had done previously.[80])

In his presidential address, James borrowed inspiration and authority from Janet's methodology and proposed that the topic of human energies was best studied in "actual persons" and "exemplified in individual lives," and that laboratory experimentation "can play but a small part." He expanded his own purview beyond those whose psychological energy was pathologically depleted, however, and insisted on the broader applicability of his project to society as a whole. His primary concern was not with these "formidable neurasthenic and psychasthenic conditions, with life grown into one tissue of impossibilities," but rather with the everyday experience of "imperfect vitality":

> Everyone is familiar with the phenomenon of feeling more or less alive on different days. Everyone knows on any given day that there are energies slumbering in him which the incitements of that day do not call forth, but which he might display if these were greater. . . . Compared with what we ought to be, we are only half-awake. Our fires are damped, our drafts are checked.[81]

The phenomenon of "second wind," James asserted, reveals to us the existence of these "reservoirs of energy that habitually are not tapped." Such experiences apply to mental as well as physical exertion, he insisted, and in some cases pushing beyond the "extremity of fatigue-distress" unleashes "amounts of ease and power that we never dreamed ourselves to own." Although habitually we fail to keep going, he maintained that when we do it is owing to either an unusual stimulus that fills us with "emotional excitement" or an idea of necessity that induces us to make the extra effort of will; "*Excitements, ideas, and efforts,* in a word, are what carry us over the dam." Of all the "classic" emotional stimulants, which James listed as "love, anger, crowd-contagion, or despair," it was anger that appeared to intrigue him the most.[82]

James had a long-standing interest in the energizing potential of angry emotions. In his *Talks to Teachers*, he discussed how lessons may harness the rousing nature of "pugnacious excitement" and the "fighting impulse" to their students' advantage; and in his essay "Is Life Worth Living?" he suggested that such incitements may be helpful to the would-be suicide in their personal struggle to find a reason to keep going: "for where the loving and admiring impulses are dead, the hating and fighting impulses will still respond to fit appeals. This evil which we feel so deeply is something that we can also help to overthrow."[83] Even within his lectures on *The Varieties of Religious Experience*, James allowed himself to be diverted, briefly, from examples of specifically religious regeneration and onto this topic. There he stated that "one mode of emotional excitability is exceedingly important in the composition of the energetic character, from its peculiarly destructive power over inhibitions. I mean what in its lower form is mere irascibility, susceptibility to wrath, the fighting temper." He positioned "inhibition-quenching fury" as a powerful solution to the pathological inertia of the divided self, with its "inner friction and nervous waste."[84]

In his Gifford Lectures James only mentioned this observation in passing, but there is evidence that he intended to follow it up at length. In the period between 1907 and 1909 he began browsing military biographies for source material that he planned to utilize in a study, which he sometimes referred to as the "Psychology of Jingoism" and at other times as the "Varieties of Military Experience." His son Henry mentioned this research in his edition of his father's letters, describing it as a "rather vague project." However, he stated that it was "safe to reckon" that, had his father lived to complete it, "two remarkable papers—the 'Energies of Men' (written in 1906) and the 'Moral Equivalent of War' (written in 1909)—would have appeared related to this study."[85]

In the latter essay, "The Moral Equivalent of War," James proposed that although the bloodshed and destruction of war are unnecessary and undesirable, the human race has need of the "capacity of heroism" that underpins all military campaigns. He respected the militarists' defense of war as the "supreme theatre of human strenuousness" and argued that "there is a type of military character which everyone feels that the race should never cease to breed, for everyone is sensitive to its superiority." Essentially, James acknowledged that warfare keeps the human race in shape, in that "militarism is the great preserver of our ideals of hardihood." He then went on to suggest alternative projects, other than international combat, which might retain these ideals and provide a different but equivalent kind of "stimulus . . . for awakening the higher ranges of men's spiritual energy." James proposed that there should be "instead of military conscription a conscription of the whole youthful population to form for a certain number of years a part of the army enlisted against *nature*. . . . To coal and iron mines, to freight trains, to fishing fleets . . . would our gilded youths be drafted off."[86]

James appears to have drawn inspiration for his "warfare against nature" solution from both his own experiences as a parent and his familiarity with the contemporary medical discourse. In 1899 he dispatched his two eldest sons off to summer jobs as forestry workers and commented approvingly in one paternal letter: "I read from what you don't say more than from what you do say, that the forestry experience is not a holiday picnic altogether, but a soldiers campaign, and you are putting it through with a stiff upper lip, resolved not to squeal or be a baby."[87] In the American neurological literature, meanwhile, the health benefits of this kind of manual labor were much lauded.

Silas Weir Mitchell (1829–1914), for example, in his text *Wear and Tear*, addressed the various aspects of American life that strain the nervous system. These, he claimed, pertain especially to the upper classes and included the energy-depleting demands of "overeducation" and excessive mental work. In contrast, he highlighted the energizing properties of an "out-door life" lived by those who "win bodily subsistence at first hand from the earth and waters . . . [in] a sturdy contest with Nature." He concluded that "some such return to the earth for the means of life is what gives vigor and developing power . . . [to a] race." Mitchell asserted that manual labor, such as that proposed by James's scheme, builds up a "capital of vitality" for the laborer and their descendants.[88]

In "The Moral Equivalent of War" James, similarly, commended the toughening effects of hard labor. In comparison, he insisted, those who live in an "ease-economy" are liable to become "soft and squeamish." At this late stage in his career, however, he did not couch his concerns in terms of the loss of muscular and vascular strength, as he had done previously. Instead his polemic dwelled on the dangers of cowardice. Here James celebrated *moral* fiber, rather than *muscle* fiber, and he implied that the health of humankind may be measured by its powers of endurance. From his perspective, the "energies and hardihoods" displayed by soldiers and laborers alike were characterized by the quality of resilience in the face of discomfort, pain, and death.[89] In *The Varieties of Religious Experience* he explained that

war and adventure assuredly keep all who engage in them from treating themselves too tenderly. They demand such incredible efforts, depth beyond depth of exertion, both in degree and in duration, that the whole scale of motivation alters. Discomfort and annoyance, hunger and wet, pain and cold, squalor and filth, cease to have any deterrent operation whatever. Death turns into a commonplace matter, and its usual power to check our action vanishes. With the annulling of these customary inhibitions, ranges of new energy are set free, and life seems cast upon a higher plane of power.[90]

In "The Energies of Men" James made this point succinctly. He pronounced that wars were "the great revealers of what men and women are able to do and bear," and he developed this line of thought with reference to the example of a Colonel Baird Smith whose actions were decisive in the 1857 siege of Delhi, in India. He described how Baird Smith, despite having had an attack of "Camp Scurvy," a foot and ankle that were a "black mass" of infection from a shrapnel wound, a suspected broken arm, and "a constant diarrhoea," managed to continue to discharge his officer duties. He concluded that "such experiences show how profound is the alteration in the manner in which, under excitement, our organism will sometimes perform its physiological work." Under more usual circumstances, James asserted "the normal opener of deeper and deeper levels of energy is the will."[91]

In the next section of his address, James explored the possibility of enhancing and refining these volitional efforts. He made the case that it is with this end in mind, of strengthening the power of the will, that "the best practical knowers of the human soul have invented the thing known as methodical ascetic discipline."[92] In his earlier publications, the practice of self-discipline had occupied an elevated status as an essential component of mental hygiene, but by the time James wrote *The Varieties* he had somewhat lost his faith in the exercise of the will as a route to health and stamina, and in this text asceticism was celebrated for its symbolic significance, which will be discussed in chapter 5, rather than its practical efficacy. His subsequent reversion to his original position would seem to have been motivated by an exchange of letters, beginning in 1905, with a Polish friend and philosopher, Wincenty Lutoslawski (1863–1954).

James featured this correspondence in "The Energies of Men," where he described Lutoslawski, albeit anonymously, as "an extraordinarily gifted man" with "an unstable nervous system" who suffered from regular and intractable bouts of extreme exhaustion. He reported that his friend had tried and failed to obtain relief for his condition from "the best specialists in Europe" before turning to hatha yoga. The regime he followed of voluntary starvation, reduced sleep, body postures, and breathing exercises had proved to be a revelation. His debilitating lethargy had disappeared, and he discovered a newfound power over his body.[93]

The letter James sent, on hearing about his friend's recovery, reveals how much it would have meant to him to find a way to improve his own flagging energy levels. It also contains the theoretical insights that came to structure one section of his talk:

> What impresses me most in your narrative is the obstinate strength of will shown by yourself . . . in your methodical abstentions and exercises. . . . I find, when my general energy is [taken up] by hard lecturing and other professional work, that then particularly what little *ascetic* energy I have has to be re-mitted, because the exertion of inhibitory and stimulative will required increases my general fatigue instead of "tonifying" me. But your sober experience gives me new hopes. Your whole narrative suggests in me the wonder whether the Yoga discipline may not be after all in all its phases simply a methodical way of *waking up deeper levels of will-power than are habitually used*, and thereby increasing the individual's vital tone and energy.[94]

In his address, James cited his friend's interpretation of his cure. He recounted Lutoslawski's conclusion that "everybody who is able to concentrate thought and will, and to eliminate superfluous emotions, sooner or later becomes a master of his body and can overcome every kind of illness. This is the truth at the bottom of all mind-cures. Our thoughts have a plastic power over the body."[95] James, in turn, refashioned this explanation in the language of contemporary psychotherapeutics:

> Suggestion, especially under hypnosis, is now universally recognized as a means, exceptionally successful in certain persons, of concentrating consciousness, and, in others, of influencing their bodies' states. It throws into gear energies of imagination, of will, and of mental influence over physiological processes, that usually lie dormant, and that can only be thrown into gear at all in chosen subjects. It is, in short, dynamogenic; and the cheapest terms in which to deal with our amateur Yogi's experience is to call it auto-suggestive.[96]

The case studies of Lutoslawski, Colonel Baird Smith, and others featured in "The Energies of Men" are notable for their common theme. The energies discovered are put to use in a triumph of human agency over evil, either the eradication of illness or incredible tenacity in the presence of pain. In his concluding thoughts, James discussed the implications of these inspirational accounts and proposed nothing less than a reimagining of the entire project of psychology, a medically minded re-

imagining that was inspired by the work of Italian pragmatist Giovanni
Papini (1881–1956), whom James had met recently for the first time, in
Rome.

James described how Papini had adopted a new conception of philosophy: "He calls it the *doctrine of action* in the widest sense, the study
of all human powers and means."[97] James suggested to his learned audience that they should follow Papini's lead in making two problems the
subject of their urgent attention: first, the need to ascertain the extent
of humankind's reserves of energy or power and, second, the matter of
how best to access these resources:

> We ought to somehow get a topographic survey made of the limits
> of human power in every conceivable direction, something like an
> ophthalmologist's chart of the limits of the human field of vision . . .
> [and] then construct a methodical inventory of the paths of access,
> . . . differing with the diverse types of individual, to the different kinds
> of power.[98]

James described his plan as "a program of concrete individual psychology" that pointed to "practical issues superior in importance to anything we know." The earlier content of his speech implied that these
practical issues comprised James's abiding interest in health and rejuvenation, and this inference is supported by a letter that James sent to
another pragmatic philosopher and friend, Ferdinand Canning Scott
Schiller (1864–1937), shortly after becoming acquainted with Papini's
work. There he confided,

> What I really want to write about is Papini. . . . I must be very damp
> powder, slow to burn, . . . for I confess that it is only after reading these
> things (in spite of all that you have written to the same effect, . . .) that
> I seem to have grasped the full import for life and regeneration, the
> *great* perspective of the program, and the renovating character for *all
> things* of Humanism.[99]

Schiller referred to his own species of pragmatic philosophy as a form
of "humanism," in recognition of his belief in the fundamental role that
human interests play in shaping truth and reality.[100] James, it would
seem, only realized certain implications of this human-centered vision
on reading Papini's writings for the first time. A review he published of
his Italian colleague's work two months later in the *Journal of Philosophy* indicated the nature of those revelations. James's review discussed
the relationship between humankind and the reality it inhabits, and he

explained that Papini's philosophy promotes "the bringing of our spiritual powers into use, and the need of making the world over instead of merely standing by and contemplating it." James conveyed Papini's accent on humanity's ability, even duty, to mold and change the world into a more ideal form.[101] He described his conviction that

> philosophy can resolve itself into a comparative discussion of all the possible programs for man's life when man is once for all regarded as a creative being. As such, man becomes a kind of god, and where are we to draw his limits? . . . The unexplored powers and relations of man, both physical and mental, are certainly enormous; why should we impose limits on them *a priori*?[102]

In the background of James's reflections lurked an emphasis on environmental adaptation that had long informed his thinking.[103] Although he had, for many years, argued vociferously against most aspects of Spencer's philosophy, in the domain of health, James had previously maintained a certain reverence for his evolutionary notions.[104] His *Talks to Teachers* evidenced his commitment to the idea that, in terms of physical well-being, the ability to adjust to one's surroundings was imperative. In this context, the external world set both the standard and the limit of human achievement; it was the arbiter of physiological perfection.

In this light it is possible to grasp what James found so appealing about Papini's philosophy. Rather than slavishly capitulate to "the world that one believes and lives and moves in," Papini preached that we should act as "man-gods" and refashion this world anew in the form that we desire. Such a bold doctrine of melioration turned the Spencerian definition of health on its head: we have no need to adapt to the environment; rather, it must be adapted to fit us and our ideals. Health and regeneration were no longer defined or confined by the demands of an external reality but by the undetermined extent of humankind's own inner resources. In a letter to the yoga-convert Lutoslawski, James extolled his newfound confidence in the potency of humanity: "I believe with you, fully, that the so-called 'normal man' of commerce, so to speak, the healthy philistine, is a mere extract from the potentially realizable individual whom he represents, and that we all have reservoirs of life to draw upon, of which we do not dream."[105]

Papini's man-gods represented one more spinning of the Jamesian web linking morality, religion, the evil of pain, and the good of health. This nexus of ideas remained a constant preoccupation for James throughout his life, and although he reconfigured it numerous times, it was bound, in every instance, by the linked concepts of energy and en-

durance. James's earliest notions of the moral universe invoked a remote
and authoritarian God of ethical expectations and acceded equal authority to the physiologists' laws of life. Self-help and perseverance, rather
than divine deliverance, were the order of the day. In this worldview,
human energies were modest and finite and to be managed carefully via
the pious application of hygienic habits. In texts such as the *Principles of
Psychology* and *Talks to Teachers*, self-discipline itself was a habit to be
established and maintained in service to the ideals of a healthy, efficient
nervous system and a useful life.

In contrast, the religious perspective that James presented in *The Varieties of Religious Experience* acknowledged the limitations of this stoic
attitude in circumstances of serious illness and despair and lauded the
energizing benefits of renouncing one's own will before the will of God.
The Jamesian faith of these lectures concerned itself not merely with the
world's moral meaning but also with the configuration of its physical
body and, in particular, the remedying of its morbid manifestations. This
Jamesian God operated interpersonally, responding directly to calls for
help and, on occasion, defying the "scientific" strictures of the natural
world and delivering us from evil with the aid of seemingly miraculous
superhuman activity.

In "The Energies of Men," the distance between God and man was
dissolved further still. There James emphasized that extraordinary powers may be accessed via a number of different routes, only one of which
involved surrendering to a celestial agency. Inspired by the prodigious
healing attained by Lutoslawski, James proposed that an individual's
will, developed via sustained feats of ascetic endurance, was an equally
effective path to these deeper sources of energy: "amounts of ease and
power that we never dreamed ourselves to own." In this way, roused by
Papini's audacious philosophical program, James advocated a faith in
the godlike powers of humanity, a society where some have the capacity to save themselves from even the most persistent forms of natural
evil. These man-gods, moreover, may disregard the biological mandate
altogether. No longer constrained by an environmental demarcation of
health they are free to define the terms of their own salvation.

CHAPTER 5

Politics and Pathology

The Political James

Over the years there have been many attempts to establish James's political sympathies. The results have been far from consistent; he has been depicted as an ally of everything from anarchism to modern industrialism, communitarian liberalism, and petty producerism.[1] The diverse nature of these claims may be attributed, in part, to the paucity of sources on which they are founded. For those wishing to reconstruct James's engagement with issues of governance and socioeconomics, there are extremely slim pickings within his surviving writings. An item from his private correspondence implies that this is no accidental feature of the historical record. In 1901 James wrote to the American reformer Ernest Howard Crosby that he wasn't sure he could see the point of actively contesting the "capitalistic social system," not because he wholeheartedly supported it but, because "with *any* system the '*Uebermenschen*' would be ingenious to get on top and *exploiter* mankind in the interests of their egoism."[2] Similarly, his most extensive public discussion of the labor question in his 1899 address "What Makes a Life Significant?" amounted to a few, not very sophisticated, paragraphs. James concluded this analysis with the following words:

Society has . . . undoubtedly got to pass towards some new and better equilibrium and the distribution of wealth has doubtless slowly got to change: such changes have always happened and will happen to the end of time. But if, after all I have said, any of you expect they will make any *genuine vital difference* on a large scale, to the lives of our descendants, you will have missed the significance of my entire lecture. The solid meaning of life is always the same eternal thing, — the marriage namely, of some unhabitual ideal, however special, with some fidelity, courage, and endurance; with some man's or woman's pains.[3]

One reading of these contributions would be to dismiss the idea of James as a politically motivated thinker altogether. There is little, certainly, to indicate that he had any sustained interest in debating the merits or otherwise of capitalism and its alternatives. Nor, it's fair to add, did he display a significant concern for the social complexities surrounding either gender or race. But to oust James from the political arena on these grounds would be to operate within an extremely narrow definition of the political. If instead a more expansive understanding of political activity is adopted, one that includes attempts to critique and reform all kinds of social structures, hierarchies, and discrimination, an important theme becomes visible in James's later work, one that has previously been overlooked.

In the excerpt above, James alluded to this theme when he answered the title question of his talk "What Makes a Life Significant?" with an emphasis on the components of endurance and pain. During the address he drew attention to the physical suffering inherent in the life of a laborer: "the backache," "dangers," and "scars." However, he stressed that he was making a broader point, one that may apply equally to anyone, regardless of their material circumstances, including the "educated man" in his comfortable home. James's aim was to convince the audience that a meaningful career should not be assessed according to conventional templates of success. He determined, instead, that faithfulness and resilience in the face of suffering are the essence of a worthy life.

In other writings from the last two decades of his life James explored and developed this message with respect to one particular group for whom suffering is a constant and inescapable part of everyday life, namely, those incapacitated by illness and infirmity.[4] James promoted the interests and status of those classed as invalids, a social category deemed quite literally in-valid by many prominent voices within his late nineteenth-century world.[5] This polemical thread, which is explicit in several of James's writings, engages with the politics of abnormality, illness, and incapacity, the politics of pathology.[6]

The biographical import of this project is unmistakable. James, as we have already seen, identified as some sort of invalid throughout his life. In his youth he perceived himself to be a sickly, useless burden on society. And even later on, as a gainfully employed university professor, he continued to berate his chronic ill health. A comment in a letter to the philosopher Charles Renouvier, sent in 1882, is typical. In it James confided that "my poor physical condition still restricts my work to the point of making me sometimes feel quite desperate."[7] In another letter, to the philosopher Francis Herbert Bradley, which he sent thirteen years later in 1895, he described himself as an "abominable neurasthenic, dogged through life by a constant sense of the impossibility of every task, and of my own impotence."[8]

The nature of these remarks suggests that James continued, throughout adulthood, to assess the value of his life in the context of how useful he felt it to be. Despite his youthful attempts to turn to "the moral life" and "measure" himself by what he strived for and not by what he achieved, he continued to worry about his level of productivity.[9] James equated ill health with a lack of results, frustration, and a loss of self-worth. In short, he believed, or more accurately feared, that the invalid was an impotent and useless creature. As the nineteenth century drew to a close, moreover, circumstances conspired to intensify James's concerns. During this period his mental health took a distinct turn for the worse. Surviving correspondence from the mid-1890s makes it possible to chronicle James's descent into a "new kind of melancholy" of the "profoundest" sort.[10]

When James set off with Alice and their children for his sabbatical tour of Europe in the summer of 1892, he was confident that his "brain fag," (mental fatigue), which he attributed to overwork and a "neurasthenic diathesis," would lift in due course. After long hours of relentless work, proofreading the pages of his newly completed *Principles of Psychology*, James was exhausted and badly in need of a holiday. By July 1893, however, after more than a year traveling with his family, it had become clear that the hoped-for rejuvenation had failed to materialize. In a letter from London he wrote to his Harvard colleague Hugo Münsterburg that he was "getting into a rather melancholic state." In the same month he wrote again, this time to his close friend Frederic Myers, about a planned trip to Switzerland: "I hope the black cloud that has weighed on my spirits for a month past may there clear off—certainly in London it remains!" But despite his hopes, his low mood persisted and back home in Cambridge, a few months later, he confessed that the new academic year's lectures seemed like a "ghastly farce" and teaching them a formidable task. Bereft of inspiration and energy he implied that he feared his creative years were over, and "the experience brings startlingly near to one the wild desert of old-age which lies ahead."[11]

It is perhaps not surprising that James should have fallen prey again to a pernicious melancholy. In many ways his circumstances mirrored those of the late 1860s, when he had previously suffered in this way. On both occasions his professional path forward lay undecided. The writing of his psychology textbook had given a structure to the last twelve

years of his career. Now that it was finished James had to find a new professional purpose to his life. Drained by the efforts he had poured into
completing the text and unmotivated by what he saw as "the stubble
of psychology," he yearned to start again in a new discipline.[12] He was
drawn to "erkentnisstheorie" (epistemology), or cosmology, but at the
same time he felt a "curious sense of incapacity" and bemoaned his lack
of original ideas.[13] It is not clear what conclusions James came to about
the underlying cause of his "melancholic state," but letters indicate that
he sought treatment from a local mind-curer whom he subsequently
credited with his cure.

The next time James wrote of his health was in correspondence from
December 1893 with his brother Henry and friends Myers and Théodore
Flournoy. By this point the worst was over. He admitted that he had
been suffering from the "profoundest melancholy from which I gradually emerged and am now all right again."[14] The timing of this experience, however, was significant. Despite his eventual recovery this episode served to reinforce James's conception of himself as a psychological
invalid at precisely the moment when such diagnoses were becoming
fraught with political consequences. Increasingly, mental disorders were
being portrayed as a burden not merely to those who suffered from them
but also to society at large. Nothing less than the decline of the human
race was at stake.

Defending the Degenerate

The idea that humanity trembled on the precipice of deterioration and
disaster was integral to nineteenth-century theories of degeneration.
Rather than a single, coherent doctrine the concept of degeneration
comprised a constellation of ideas, fears, and conjectures, which originated from within the medical community. It was a model of disease,
with a psychiatric emphasis, that rested heavily on the notion of heredity. Sufferers were born and not made, tainted from birth with the blood
of degenerate stock. Furthermore, as a degenerate family continued to
reproduce, their offspring would become progressively more incapacitated resulting, ultimately, in an increased incidence of idiocy and steril-
ity. Although these theories had been around since the 1850s, during the
mid-1890s they became a much more vociferous presence, both within
the medical literature and in the popular press.[15]

One particularly noteworthy contribution to this public discourse
was a book, first published in America in 1895, by the Austrian author
and physician Max Nordau (1849–1923). His text *Degeneration* was
deemed by one American reviewer to be "one of the most widely-read

books of the season," a book that seemed to have engrossed "the attention of the talking public, at least for a short time, to the exclusion of
everything else."[16] Nordau's work took ideas about the degenerate, and
in particular the mental characteristics, or "stigmata," that were considered to mark them out from the rest of the population, and applied
them to various contemporary artists and writers and in some cases to
entire schools of thought.[17] His diagnoses were damning. From the Pre-
Raphaelites to Tolstoy, Wagner, Ibsen, and the school of Zola he found
evidence of degenerate minds at work. All of these examples of "odd aesthetic fashions," "realism or naturalism, 'decadentism,' neo-mysticism,
and their sub-varieties, are manifestations of degeneration and hysteria, and identical with the mental stigmata which the observations of
clinicists have unquestionably established as belonging to these," he insisted.[18]

Nordau's book was more than an attack on contemporary aesthetic
trends, since the painters, poets, and musicians on whom he focused
were portrayed as representatives of a wider "mental epidemic" of degeneracy and hysteria within society. Whereas formerly such conditions
had manifested themselves only sporadically and had "no importance in
the life of the whole community," the present age with its multitude of
fatiguing discoveries and innovations had, he argued, weakened a whole
generation and left them susceptible to these maladies, which had become a "danger to civilization."[19]

This depiction, of widespread nervous exhaustion brought on by the
trappings of life in the modern fin de siècle world, invoked the rhetoric previously associated with James's self-diagnosis of choice, neurasthenia. In addition, Nordau considered degeneracy to be the province,
primarily, of the "upper stratum of the population of large towns," precisely those "brain-workers" considered by American neurologists such
as George Miller Beard and Silas Weir Mitchell to be most likely to develop neurasthenic symptoms.[20] And indeed, Nordau went so far as to
make this link, between the degenerate and the neurasthenic, explicit.
In one of the opening chapters of his book he stated categorically that
neurasthenia was a "minor stage" of degeneracy. He was not the only author at this time to bring neurasthenia inside the degeneration debate,
and this move, which constituted a break from the earlier degeneration
theorists' work, did not go uncontested.[21]

An American reviewer, neurologist Joseph Collins (1866–1950), took
issue with this particular aspect of Nordau's thesis in his review for the
Journal of Nervous and Mental Disease. He opined, "for the sake of our
overworked friends we would demur to neurasthenia being called a minor stage of degeneracy or a forerunner of hysteria [Nordau also allied

hysteria with degeneracy], for neurasthenia is no more a minor stage of hysteria than rheumatism is a minor stage of gout."[22] The crucial, unspoken subtext to his comment was the relative social standing of a diagnosis of neurasthenia versus one of degeneracy.

In its original Beardian incarnation neurasthenia was the hallmark of a respected, hard-working, professional man, someone who lived in one of the world's most advanced, and thus exhausting, civilizations (America), and consequently had to pay the price with his health. In essence, the locus of this disease lay within society itself, rather than the sufferer. A diagnosis of neurasthenia offered the afflicted the opportunity to present their symptoms to their confreres in a socially acceptable form. Beard referred to it as a "distinguished malady," and James himself remarked, in a letter to a friend, that the term neurasthenia was a "felicitous . . . invention."[23] It is notable, moreover, that he freely described himself as neurasthenic to a wide variety of correspondents. The notion of an underlying hereditary predisposition toward the condition was touched upon by Beard but quickly passed over, and he implied that it was of little significance.[24]

The degenerate, on the other hand, was commonly depicted as the living embodiment of a savage, uncivilized past and the atavistic portent of future racial decline. In their bodies lay the seeds of society's downfall. Nordau, for example, referred to the degenerate as "anti-social vermin."[25] In this way, Nordau's and Beard's portrayal of the neurasthenic were poles apart: "Whereas the American neurasthenic was only a short step away from complete cure, the European neurasthenic was a few generations away from the extinction of the race."[26] In short, as the century drew to a close, James's own diagnosis of his condition was transformed, at the hands of some of the degeneration authors, from a badge of honor into a shameful, ingrained stain. In James's own words, to call a man "a 'degenerate'" was to group him "with the most loathsome specimens of the race."[27]

The social stigma attached to degeneracy was by no means its most disturbing aspect. James, as discussed earlier, had married only after extensive deliberations. The chance that he might pass on any aspect of his ill health to his progeny appears to have been a very real and very painful prospect for him at the time, and there is evidence that these concerns never really left him. In 1882, an ex-Harvard student wrote to James for advice about the hereditability of "nervous or mental disease" and his prospective marriage partner. James apprised him of medical opinion on the subject but acknowledged that such personal decisions could be ethically "delicate," regarding "the question of duty to offspring and society," and "will be one of the great moral 'crosses' which our posterity

5.1 James photographed with his daughter, Margaret Mary James (Peggy), in March 1892. (Letters to William James from various correspondents and photograph album [MS Am 1092 (1185), 3, Box: 12], Houghton Library, Harvard University.)

will have to bear."[28] By the time Nordau et al. were exciting debate in the mid-1890s, James was a father to four children. The possibility that his own neurasthenic symptoms were indicative of a grave physiological inheritance would, most likely, have greatly alarmed James.[29] This supposition is supported by the fact that, several years later, his wife, Alice, attempted to keep the knowledge of their daughter Peggy's first "nervous breakdown" from her husband.[30]

Worse still, James discerned a militant faction operating under the auspices of the degeneration doctrines. In his own review of Nordau's text, for example, he described the author, only half-jokingly, as someone who seemed to harbor "delusions about a conspiracy of hysterics and degenerates menacing the moral world with destruction unless the sound-minded speedily arm and organize in its defence."[31] In Nordau's own words, society was languishing under a "sort of black death of degeneration and hysteria." In such circumstances, he insisted, "it is the sacred duty of all healthy and moral men to take part in the work of protecting and saving those who are not already too deeply diseased." Degenerates are "enemies to society of the direst kind," and, accordingly, it is the responsibility of the healthy to "crush under his thumb" these "anti-social vermin."[32] This medical-material extremism appeared at a time when James was especially vulnerable, having only recently recovered from the protracted episode of melancholy that had severely shaken him. Significantly, it is at this moment that he began to explicitly, and publicly, challenge the perception and position of the psychological invalid within society. James's concern, as he expressed it, was that medical diagnoses had become "clubs to knock men down with."[33]

Harvard class records indicate that, despite maintaining a keen interest in mental pathology since his early adulthood, James only commenced teaching on this topic in 1893, immediately after he returned from his European sabbatical and following the publication of the original German edition of Nordau's text.[34] He continued with these courses on "abnormal psychology" until his leave of absence, due to his heart condition, in 1899. Although no notes connected specifically with these student classes have come to light, James also lectured on this subject to the general public, and a manuscript version of these talks has survived. James delivered them, in one form or another, on at least three separate occasions including, most notably, at the Lowell Institute in 1896, where he dealt with the following eight topics: "Dreams and Hypnotism," "Hysteria," "Automatisms," "Multiple Personality," "Demoniacal Possession," "Witchcraft," "Degeneration," and "Genius."[35] He referred to them as his lectures on "Exceptional Mental States," and this title, as will become clear below, was indicative of their principal objective. These lectures constituted a pointed social commentary on the politics of psychopathology.

Both the opening thoughts of the first talk in the series and the closing words of the final hour focused explicitly on this theme. The first lecture was entitled "Dreams and Hypnotism," and although some parts of his notes were written out more fully, in prose form, this section survives only as a list of disjointed phrases: James's cues. It began with the following reflections:

> Common distinction: "healthy" & "morbid"
> Can't make it sharp.
> No one thing is morbid.[36]

From these notes, although they are brief, it is possible to infer James's general message. His intention, it would seem, was to plant a seed of doubt in the minds of his audience as to the easy familiarity with which they used the linked concepts of "health" and "morbidity." He also hinted at two separate lines of reasoning, which illustrated his point in different ways. As the lectures progressed James developed both these arguments further.

The first rested on the assertion that it is very difficult to draw a line of demarcation between the two, supposedly opposite, states of psychological health and psychological morbidity. James argued that characteristics that are regarded as normal or healthy shade off, gradually, into conduct that is deemed to be abnormal. Over the course of the lectures he took pains to elaborate and reiterate this proposition with illustrative examples. In the lecture that specifically addressed degeneration, he discussed certain types of behavior that were considered to be the giveaway stigmata of a degenerate mind. These included obsessive ideas and impulses that are outside of one's control. James pointed out that there is a "normal germ" of such behaviors in many people when, for example, they feel the need to check that they have actually sealed a letter, turned off the gas, or locked the door, despite having already carried out these actions.[37] The only difference between these everyday experiences and those that are labeled pathological in the psychiatric literature is, James asserted, that in the "morbid cases" the compulsion to check returns again and again in an insistent form.

In this same lecture, he also observed that neurasthenic "phobias," such as the fear of open spaces, of enclosure, crowds, solitude, disease, contamination, or just of "something dreadful" are, again, present in "germ" form in most people when they are "unnerved."[38] Using these and other examples James repeated, at several junctures, the idea that normal and abnormal behaviors are only distinguished by degree of extremity, rather than by any qualitative difference.

This general idea that pathological and normal phenomena are essentially the same and only vary in intensity was nothing new and certainly not an original Jamesian standpoint. It was a commonplace observation

in the medical texts of this time. And indeed, the psychological and psychiatric communities were among the most enthusiastic proselytizers
of the theory.[39] But James was making a distinctive point. The crucial
difference was the ultimate end to which he mobilized this assertion.
Colleagues, such as the French psychologist Theodule Ribot, used it
to justify their specific research methodology. Ribot, Pierre Janet, and
others argued that as pathological phenomena were simply exaggerated
manifestations of normal phenomena, they offered the perfect opportunity for studying the human mind at work; in effect, disease amplified
normal, imperceptible mental states and made them more visible to the
investigator. James, however, was primarily concerned with a very different endgame. He was intent on renegotiating the reputation of those
diagnosed with various forms of mental pathology.

According to one of James's students during the 1890s, Dickinson
Sergeant Miller (1868–1963), this was a central theme of his teaching.
In his account of James's course on abnormal psychology, he recalled a
class visit to an insane asylum, where they had seen a "dangerous, almost
naked maniac." Miller recounted James's comment that "President Eliot would not like to admit that no sharp line could be drawn between
himself and the men we have just seen, but it is true."[40] The reference to
the then president of Harvard University Charles William Eliot (1834–
1926) was presumably not intended as a personal slight. Instead, James
was trying to narrow the gap between two people who would ordinarily
be assumed to inhabit different ends of the social and psychiatric spectrum, namely, a distinguished leader and a naked madman. In the years
that followed, moreover, Eliot became a vocal proponent of the American eugenics movement. He supported legislative restrictions on the
marriage of those "burdened with inheritable weaknesses" and called
for the "compulsory seclusion of all defectives."[41] It seems likely that
his sympathies with such causes were already in evidence at this time,
prompting James's provocative comment.

Over the course of his "Exceptional Mental States" lectures, James
also constructed a second line of argument contesting the validity of
the degeneration diagnoses and their social implications. Specifically,
James opposed the notion that a single element of abnormal psychology should be decisive, insisting that "no *one* thing can make a man
unsound," and established, instead, a less pessimistic and broader definition of mental health, stating that it is an "affair of balance."[42] James
expounded these ideas very clearly in a book review that he published,
in 1895, one year before his Lowell Lectures. The review in question was
part of a linked set of four, all dealing with books focusing on degener-

ation, which James wrote for the journal *Psychological Review*.[43] These reviews are useful as they contain material that seems to map directly onto passages of his later talks and is otherwise only available in note form. The last in the group was of William Hirsch's book on genius and degeneration. This text was the one most favorably received by James.

In private discussions with his publisher, Henry Holt, James referred to Hirsch's *Genius and Degeneration* as "a capital book, full of good sense and really instructive."[44] This praise was undoubtedly borne of Hirsch's critical appraisal of Nordau. Hirsch's work was frequently characterized, as the reviewer in the *Journal of Nervous and Mental Disease* put it, as the "answer, or refutation" to Nordau's text.[45] Hirsch directly contradicted Nordau's assertion that the present age was one blighted by epidemics of mental diseases such as hysteria. In comparison with earlier ages, he insisted, the current period was, if anything, healthier in this respect, and, he concluded, there is "no proof of the alleged universal degeneration in the highly civilized nations."[46] In general, Hirsch's book displayed an "optimistic" tone at odds with the other publications on the topic of degeneration.[47] James made his own allegiances very clear, declaring publicly that it "would be well if all the admirers of Lombroso, Nisbet, and Nordau could be compelled to read Dr. Hirsch's admirable study."[48] At the end of his review of the book, he summed up precisely what he objected to in the rest of the degeneration literature:

> There is a strong tendency among these pathological writers to represent the line of mental health as a very narrow crack, which one must tread with bated breath, between foul fiends on the one side and gulfs of despond on the other. Now health is a term of subjective appreciation, not of objective description. . . . There is no purely objective standard of sound health. Any peculiarity that is of use to man is a point of soundness in him, and what makes a man sound for one function may make him unsound for another.[49]

James's comments were in part a criticism but also an attempt to transpose the prevailing, negative definition of mental health, based on the absence of all pathology, into a more inclusive, positive definition constructed around a broad concept of usefulness to society. Within this alternative paradigm any personal characteristic that, in certain circumstances, proves "useful to our kind" may, "in that context," be considered healthy. To emphasize the fallaciousness of the normative project of the degeneration authors, James described the features of someone whom these writers would hold up as a perfect specimen of humanity:

The only sort of being, who can remain as the typical normal man, after all the individuals with degenerative symptoms have been rejected, must be a perfect nullity. . . . Being free from all the excesses and superfluities that make Man's life interesting, without love, poetry, art, religion, . . . he is the human counterpart of that "temperance" hotel of which the traveler's handbook said: "It possesses no other quality to recommend it." . . . Few more profitless members of the human race can be found.[50]

After dismissing the celebrated "normal man," James then proceeded to highlight the potential of those deemed degenerate. This section of his argument was developed more fully in his notes for his lecture on genius. There he referred to his earlier talk on degeneration and the characteristics that were regarded as typical of those with a "psychopathic temperament." He explained that the mental "stigmata" on this list, which included "moral intensity," "obsessiveness," and a single-minded "belief in their mission," can, if they happen to be coupled with a keen intellect, lead to works of genius, important social contributions. Intellect alone, however, without this additional, supposedly pathological ingredient of "zeal and doggedness" was, he argued, much less likely to produce remarkable, world-changing results.[51]

James ended this lecture, which was the last of his series, with a request. He suggested that, bearing in mind their contributions to the weal of the human race, we should approach the abnormal regions of human nature with "a certain tolerance, a certain sympathy, a certain respect" and "above all a certain lack of fear."[52] His take-home message was that there should be a reevaluation of the role and reputation of those who display or suffer from psychopathological traits. With this in mind James had challenged, on two different fronts, contemporary understandings of mental health; he had tried to bridge the perceived divide between the sane and the insane and championed the worth of those labeled as degenerate. James's lectures on the psychology of exceptional mental states were not the only place where he disputed the assessment and social standing of the invalid, however. He also pursued this agenda, from another angle, in several of his writings on ethics and religion from this period.

Validating the Invalid

In May 1895 James gave an address to the Harvard Young Men's Christian Association entitled "Is Life Worth Living?," followed, a year later, by another address "The Will to Believe," this time to members of the Yale and Brown Philosophical Clubs. Later published as essays these two well-known texts contained ideas about the justification of moral and religious beliefs that James had, for the most part, worked out twenty or so years earlier. In "Is Life Worth Living?" whole passages were lifted, almost verbatim, from letters and notes that James had written during his early adulthood. Similarly, "The Will to Believe" included philosophical arguments that he had originally forged in this same period.[53]

There is, then, an interesting question as to why James chose the mid-1890s to, in the words of his first biographer, Perry, "reaffirm and amplify the faith of his youth." Perry asserts that James, suffering once more from episodes of "brooding melancholy," again felt the need of his "saving gospel," but this is only half the story.[54] He neglects the significance of the transition from *private* gospel to *public* preaching. In the 1890s, as opposed to the 1870s, James wasn't concerned merely to save himself, his message was intended for the rest of society. These two addresses speak to the same manifesto that motivated his talks on "Exceptional Mental States," which were first delivered in the following year, and his subsequent lectures on *The Varieties of Religious Experience*. Together, elements of all these writings constituted James's multifaceted answer to his political enemies: the gathering armies of medical materialism; the harbingers of degenerative doom.

James used his ethical and religious texts to contest the growing authority of the idea that biological superiority represents the ideal form of humanity. In its place, he sought to establish an alternative criterion, based on the concept of "spiritual heroism." This sphere of possibility was, for James, bound up with religious ideals, practices, and beliefs. It was a sphere that validated suffering and those who suffer and was within reach of even the most afflicted of society's invalids.

The contrast between these two opposing evaluations of human potential was drawn most vividly by James in his chapter on "The Value of Saintliness" in *The Varieties of Religious Experience*. There James introduced the German philosopher Friedrich Nietzsche's (1844–1900) opinion of sainthood. According to James, Nietzsche presented the saint as a "sophisticated invalid, the degenerate *par excellence*, the man of insufficient vitality. His prevalence would put the human type in danger."[55] He then went on to cite a passage, which he translated and abridged, from Nietzsche's 1887 *Zur Genealogie der Moral* (*On the Genealogy of Morality*):

> The sick are the greatest danger for the well. . . . The *morbid* are our greatest peril—not the "bad" men, not the predatory beings. Those born wrong, the miscarried, the broken—they it is, the *weakest*, who

are undermining the vitality of the race, poisoning our trust in life, and putting humanity in question.[56]

James summed up Nietzsche's sentiments as expressing a universal debate, specifically a clash of values. What is at stake, he suggested, is our concept of the "ideal" man; there is a question about whether this should be "the carnivorous-minded 'strong man'" or a man who espouses the saintly virtues of "gentleness and self-severity." These two human types, the "saint's type, and the knight's or gentleman's type, have always been rival claimants of this absolute ideality," he argued.[57]

James's own treatment of these two types implied that he too conceived of an association between saintliness and invalidism. Although he originally set out to compare "aggressiveness" and "non-resistance," and their respective validity as approaches to life, these two categories morphed silently during his analysis into "strong men" and "saints," intimating that sainthood commonly entails some degree of physical inferiority. This inference is supported by James's idiosyncratic list of "the greatest saints" in *The Varieties*. He would include, for example, those who had never been canonized by any church, figures such as William Ellery Channing, Agnes Jones, and Dora Pattison who were known for their delicate health and in the case of Jones and Pattison their deaths at a young age from breast cancer and typhus, respectively.[58] (Interestingly, though the rest of the list were known primarily for their preaching or theological leadership, these two women were famous for their charitable work nursing the sick in various hospitals and poorhouses.[59])

James also conflated strength and militancy, and while he leaned explicitly on Nietzschean notions, his own analysis was equally rooted in his immediate cultural context. In post–Civil War America, traditional ideals of manhood that stressed the utmost importance of intellectual and moral standing were displaced by an emphasis on the martial virtues of action and physical fearlessness. This valorization of bodily strength and courage was enthusiastically celebrated on the Harvard athletics field, and contemporaries of James's, including Oliver Wendell Holmes Jr. and Henry Cabot Lodge, made the case that sports, especially dangerous ones, were a worthy training ground for the combative attributes of this new masculinity and for war itself.[60]

James, meanwhile, drew implicitly on popular associations between sickness and spiritual transformation in his construction of the invalid saint as a rival to this knightly conquest of American manhood. In his nineteenth-century world, there was a widely held belief that serious or protracted ill health fostered moral betterment and religious conversion.[61] This belief was epitomized, for example, in the fictional character of Katy Carr in the novel *What Katy Did* (1872), who injured her spine and improved herself at "The School of Pain."[62]

In *The Varieties* James espoused a conception of sainthood so broad as to be within reach of anyone. He ended his chapter on "The Value of Saintliness" with the message "Let us be saints, then, if we can, whether or not we succeed visibly and temporally. But in our Father's house are many mansions, and each of us must discover for himself the kind of religion and the amount of saintship which best comports with what he believes to be his powers and feels to be his truest mission and vocation."[63] This informal and inclusive definition of saintliness, and its link with ill health, were also in evidence in James's personal correspondence. In a letter to a friend, Alice Smith, in 1900, for example, he wrote of his treatment for heart disease and of how "there's a potential healthy villain . . . inside of every invalid saint."[64]

Ultimately, in his lectures, James answered the question of whether we should favor the "healthy villain" or the "invalid saint" by challenging the evolutionary framework that underpinned his contemporaries' promotion of physical vigor. This framework was built around the notion that the ideal specimen, of any species, was one that was best adapted to its surroundings. "How is success to be absolutely measured," James asked, "when there are so many environments and so many ways of looking at the adaptation?"[65] The verdict, he pointed out, will vary according to which environment is considered.[66] Taking Saint Paul as one example, James conceded that "from the biological point of view [he] was a failure, because he was beheaded!" From the perspective of "the larger environment of history," however, his success, measured in terms of his prowess in inspiring subsequent examples of saintliness, virtues such as "humble-mindedness," "patience," and "charity," was indisputable. Similarly, the "greatest saints," or "spiritual heroes," are "successes from the outset . . . and, placed alongside of them, the strong men of this world and no other seem as dry as sticks, as hard and crude as blocks of stone or brickbats." He also argued that "a society where all were invariably aggressive would destroy itself by inner friction."[67]

In the *Varieties of Religious Experience*, James concluded that the evolutionary biologists were wrong to champion the coarse strength of brawn above all else, and that by enlarging our ordinary, taken-for-granted definitions of "adaptation" and "environment," we arrive at a different estimation of human value. He depicted saintliness as a supremely positive form of social participation that was accessible to anyone, however feeble their physiological endowments. In other texts, moreover, James went further still in his validation of the invalid and their contributions. He utilized the moral and religious domain to construct a vision of spiritual heroism that was not merely available to society's invalid citizens but uniquely suited to their experience of life.

Broadly speaking James's two addresses "The Will to Believe" and "Is Life Worth Living?" were concerned with establishing two interconnected lines of reasoning. The former expounded a defense of religious faith, in the face of the materialists' argument that such beliefs are unsupported by evidence. The latter address, meanwhile, asserted that, given this faith in the "unseen world" of religion, instances of human suffering assume a hitherto unrecognized civic significance. They are heroic acts, no less, which play a vital role in determining the fate of humankind.

This conjoining of the concept of heroism with the experience of suffering takes place in the concluding paragraphs of "Is Life Worth Living?," in the section that James referred to as the "soul of my discourse."[68] There he first defined his terms, stating that "a man's religious faith"

> means for me essentially his faith in the existence of an unseen order of some kind in which the riddles of the natural order may be found explained. In the more developed religions the natural world has always been regarded as the mere scaffolding or vestibule of a truer, more eternal world, and affirmed to be a sphere of education, trial or redemption.[69]

It is instructive to consider this Jamesian description of religious faith in the context of his many references to evil. As his writings attest, the primary "riddle of the natural order," according to James, was the "problem of evil," and the type of "evil" foremost in his mind was the misery of illness and pain. In this light, a major function of the "unseen order" upheld by religious faith was, for James it would seem, to provide an explanation for, or justification of, the anguish of those afflicted by disease.

The next part of his address strongly supports this interpretation. Here James used an analogy involving the lives of our domestic pets to illustrate our imperfect grasp of the "unseen world" and its relationship with the natural world. It is quite possible, he explained, that we currently "lie soaking in a spiritual atmosphere, a dimension of being that we at present have no organ for apprehending."[70] For example, he explained, our dogs are part of our human life but not of it, in the sense that they cannot comprehend our actions in the same way that we do. The series of events in which they play an active role harbors an inner meaning that is not accessible to their limited intelligence.[71] James asked his audience to consider a specific example from his medical school days of a dog being vivisected in the laboratory:

He lies strapped on a board and shrieking at his executioners, and to his own dark consciousness is literally in a sort of hell. He cannot see a single redeeming ray in the whole business; and yet all these diabolical-seeming events are often controlled by human intentions with which, if his poor benighted mind could only be made to catch a glimpse of them, all that is heroic in him would religiously acquiesce. Healing truth, relief to the future sufferings of beast and man, are to be bought by them. . . . Lying on his back on the board there he may be performing a function incalculably higher than any that prosperous canine life admits of.[72]

The proposition that animal and by extension human suffering may be situated as a form of heroic action is explicit within this paragraph. The dog's experience is, moreover, analogous to one very specific form of human distress, namely, the pain of bodily affliction; and the implication is that the kind of suffering James had in mind, throughout this passage, was the agony borne by those who are sick or infirm. A personal letter lends further weight to this reading.

James first employed his parable of the vivisected dog, twenty years earlier in 1875, in a letter to his brother Bob. On that occasion he began by empathizing with his brother's recent eye problems: "it's a great evil no doubt, the more so from its constitutional effect upon the spirits." James then went on to urge his brother to fight against his despondency, firing him with the same canine call to arms that he would relate in his lecture many years later. Within the context of the letter his brother's eye troubles were equated to the pain of vivisection experienced by the dog in the laboratory. James brought attention to the potential glory of his brother's situation, and he depicted a worldview in which patiently enduring the symptoms of ocular infirmity was elevated to the level of a vocation. He consoled his brother with the assurance that "you are called to a harder but also a higher lot than most of us, and if you respond it will be better than if you had not been called."[73] Several years later, in correspondence with his cousin Kitty Prince who experienced considerable mental ill health, James touched on this idea again. He referred to a supernatural, unseen world that will one day be revealed, the nature of which will bring about a reevaluation of the worth of various individuals. He wrote: "Of course I agree about what you say of the battlefields of each individual. There will be a strange transformation scene about the relative importance of various careers, on the day when the veil shall be lifted. Then indeed the last shall be first!"[74]

James's evocation of the battlefield in his letter to his cousin was not an isolated occurrence. The same metaphor also featured in his talks "Is Life Worth Living?" and "The Moral Philosopher and the Moral Life." In each instance, he deployed overtly military scenarios to summon a spiritual narrative, and he implied that those who fight metaphysical battles are due the same glory and respect as the soldiers of legend. The metaphysical war in question was the war against the evil of pain, and in his essay on "The Moral Equivalent of War," he went so far as to identify certain martial virtues in the work of priests and doctors. James took contemporary ideals of military manhood and bent them to his own political will. He inverted the association of the warrior with physical strength and opened up this hero role to invalids.[75]

James also returned to and circled around these subversive ideas about heroism in *The Varieties of Religious Experience*, in his lecture on "The Value of Saintliness," when he broached the topic of asceticism. There he took up what he acknowledged was an unfashionable position toward this kind of saintly activity. He proposed that his audience could no longer sympathize with "cruel deities" and would find "the notion that God can take delight in the spectacle of sufferings self-inflicted in his honor . . . abhorrent." He claimed, however, that the spiritual essence, if perhaps not the more lurid manifestations, of such impulses was important and relevant. In his summary of this section in the contents pages of the lectures, he stated that "asceticism symbolically stands for the heroic life," and in the text itself, he expanded on this interpretation:[76]

> For in its spiritual meaning asceticism . . . symbolizes, lamely enough no doubt, but sincerely, the belief that there is an element of real wrongness in this world, which is neither to be ignored nor evaded, but which must be squarely met and overcome by an appeal to the soul's heroic resources, and neutralized and cleansed away by suffering.[77]

Set in the wider context of his other writings, James's construction of an ascetic saint may be understood as someone who enters voluntarily into the role of sufferer in order to pay homage to the metaphysical importance of suffering and, by implication, to the invalid and their role in society. He maintained that the evils of "pain and wrong and death must be fairly met and overcome in higher excitement, or else their sting remains essentially unbroken."[78] Accordingly, those who are called upon to bear this burden deserve the accolade of hero and the admiration of society, as opposed to its fear or condemnation.

The Voice of the Sick

One of the most disturbing aspects of the contemporary medical framing of disease and the diseased, from James's perspective, was the way
that it infringed on the rights of certain groups to be heard and taken
seriously. Among the first themes he broached in the opening chapter
of *The Varieties of Religious Experience* was the widespread psychiatric
censure of mysticism. In this chapter, entitled "Religion and Neurology," James took up an epistemological debate with clear political consequences, one that centered on the question of what kind of person is
fit to reveal the truth. Specifically, he expressed his concerns about the
underhanded tactics of the medical materialists.

In this section James introduced the work of the English alienist
Henry Maudsley and other scientific figures, such as the French physician Charles Binet-Sanglé (1868–1941) who went on to question the
psychological health of Jesus in his book *La Folie de Jésus* (The madness of Jesus).[79] He generalized their observations and described how
they diagnosed mystical insights as merely symptoms of various kinds of
mental disease and, in doing so, undermined the authority and authen-
ticity of any such experience. This "simple-minded system of thought . . .
snuffs out Saint Teresa as an hysteric, Saint Francis of Assisi as an hereditary degenerate."[80]

This approach was particularly clearly exemplified in Maudsley's
book *Natural Causes and Supernatural Seemings*. In this text Maudsley
created a generic diagnosis for patients displaying the "perverted sen-
sibilities" of the typical mystic. They were not, as they claimed, com-
muning with the supernatural world, rather they were, he argued, simply
suffering from the ill effects of "psycholepsy." He insisted that an "evil
heritage" was largely to blame for such conditions, which involve the
"irregular action of one or more of the numerous and diverse nerve-
centres or nerve-tracts which make up the complex unity of the brain."[81]

In *The Varieties* James pointed out that at the center of this kind of argument lay an intellectual sleight of hand. There was, he argued, a glar-
ing lack of reflexivity inherent within such lines of reasoning. They rest
on the assumption that all mental states are dependent solely on bodily
conditions, but, he reminded his readers, according to this general physiological postulate "scientific theories are organically conditioned just
as much as religious emotions are." Unless these medical materialists
have worked out, in advance, some psychophysical theory connecting
certain sorts of physiological conditions with privileged access to truth,
he continued, they have no grounds on which to draw their conclusions

about mysticism. He summed up the injustice of their position, maintaining that

> medical materialism . . . is sure, just as every simple man is sure, that
> some states of mind are inwardly superior to others, and reveal to us
> more truth. . . . It has no physiological theory of the production of
> these its favorite states, by which it may accredit them; and its attempt
> to discredit the states which it dislikes, by vaguely associating them
> with nerves and liver, and connecting them with names connoting
> bodily affliction, is altogether illogical and inconsistent.[82]

James continued by asking his readers to be honest enough to admit
to themselves that when they think of certain states of mind as superior to others it has nothing to do with their "organic antecedents" and
everything to do with how pleasurable or useful these thoughts turn
out to be. Accordingly, he requested that they deal with the fruits of all
types of mind, morbid or otherwise, in the same way, namely, by putting them to the test.[83] Furthermore, he added, reiterating one of the
key sentiments from his "Exceptional Mental States" lectures, it is likely
that in some instances those mental characteristics that are deemed to
be pathological will prove extremely useful, and perhaps even indispensable. On this occasion he limited his argument to the domain of
metaphysical expertise.

According to his friend and fellow American philosopher Miller, the
argument that James put forward was unusual. In a letter Miller commented, having read an earlier draft of the first two lectures of *The Varieties*,

> I do not think as you seem to that your thought about mysticism
> having its roots in *disease*, etc. is a familiar one to cultivated or even
> to philosophical men; if you could even pause upon this point and
> arrive at some further *rationale* of the matter, as I seem to remember
> that you have done already in talk, it would fatten & strengthen the
> lecture.[84]

In the published version of the lecture James made his case as follows:

> No one organism can possibly yield to its owner the whole body of
> truth. Few of us are not in some way infirm, or even diseased; and
> our very infirmities help us unexpectedly. In the psychopathic tem-
> perament we have the emotionality which is the *sine qua non* of moral
> perception; we have the intensity and tendency to emphasis which

are the essence of practical moral vigor; and we have the love of metaphysics and mysticism which carry one's interests beyond the surface of the sensible world. What, then, is more natural than that this temperament should introduce one to regions of religious truth.[85]

In keeping with these observations, it is notable that, for James, the occupation of philosopher appeared to carry strong associations with infirmity and illness. In his private letters there are many lighthearted references to this effect. In 1887 he wrote to the American mystic and poet Benjamin Paul Blood, discussing Blood's career as a gymnast, fighter, and farmer: "I am so delighted to find that a metaphysician *can* be anything else than a spavined [lame or maimed], dyspeptic individual fit for no other use. Most of them have been invalids. I am one, can't sleep, can't use my eyes, can't make a decision, can't buy a horse, can't do anything that befits a man."[86] And in a letter to his friend John Jay Chapman, in 1898, on the topic of universities and their value, James ended on a jocular note: "When the state provides all sorts of asylums and institutions for incapables, why should Universities not be provided as places of retreat for philosophers like myself and Royce and others, who would otherwise go *zu Grunde* [perish] in the cruel struggle for existence?"[87]

There were also several instances within James's public writings when he made a point of referring to the health problems of a particular philosopher. In *A Pluralistic Universe*, for example, he introduced Gustav Fechner (1801–87) and detailed how "overwork, poverty, and an eye-trouble ... produced in Fechner, then about thirty-eight years old, a terrific attack of nervous prostration with painful hyperesthesia ... from which he suffered three years, cut off entirely from active life."[88] Out of the inner desperation brought on by his illness, James explained, came Fechner's professional commitment to expounding his unique brand of metaphysics. In *The Varieties* he mentioned another "venerated teacher of philosophy": a Professor Lagneau of Paris. The biographical information James gave in this case was much scantier and comprised a single detail, namely, that Lagneau was a "great invalid."[89] Another, better-known example can be found in James's edited selection of his father's literary remains. There James recounted, at length, Henry James Sr.'s mental breakdown, when he became gripped for long periods by acute pathological fear, and he described how this episode of his father's life shaped and informed his religious philosophy.[90]

In *The Varieties* James explored this link between ill health and philosophical vocation in some detail. In these lectures he divided the world's metaphysicians, amateur or otherwise, into two general types, the healthy-minded and the sick-souled. In part what separates these two groups, he established, is their stance on evil. Whereas the healthy-minded deal with evil by ignoring it, the religions of the sick-souled take evil into account. James described the essence of the sick soul's philosophy as follows:

> Now in contrast with . . . healthy-minded views . . . if we treat them as a way of deliberately minimizing evil, stands a radically opposite view, a way of maximizing evil, if you please so to call it, based on the persuasion that the evil aspects of our life are of its very essence, and that the world's meaning most comes home to us when we lay them most to heart.[91]

At this stage in the text James had yet to make his own affiliations clear, but later in the same chapter he reiterated the sentiment above and claimed this position as his own, stating that

> there is no doubt that healthy-mindedness is inadequate as a philosophical doctrine, because the evil facts which it refuses positively to account for are a genuine portion of reality; and they may after all be the best key to life's significance, and possibly the only openers of our eyes to the deepest levels of truth.[92]

Bearing in mind the close mapping of the category of evil onto the category of illness within James's worldview, these passages imply that ill health and suffering assumed, for James, some sort of epistemological rite of passage.[93] From this perspective it is only through affliction, and the accompanying pain, that we gain access to the "deepest levels of truth." Such an interpretation is supported by other sections of *The Varieties*. Indeed, James's choice of the terms "healthy-minded" and "sick-souled," to denote those who minimize and maximize evil respectively, is a case in point. He clarified, moreover, that even those who have been the victim of the most "atrocious cruelties of outward fortune" may be blind to the radical pessimistic wisdom of the sick-souled, if they do not also have a "neurotic constitution." To enter these metaphysical depths, he specified, "something more" is needed than "observation of life and reflection upon death." He stated, unequivocally, that the individual must rather "in his own person become the prey of a pathological melancholy."[94]

In summary, the metaphysical realm appeared to represent, for James, a dimension within which he could validate the invalid. Religious faith in an "unseen world" could make sense of their suffering and raise it to the level of a heroic act. On the supernatural battlefield of life, the

physically weak became society's most able soldiers; the ultimate annihilation of all that is evil lay in their hands. And, in addition, those who
endured ill health, and especially mental disorders such as melancholy,
acquired the faculty of a privileged witness. Via their experiences, they
alone were permitted to access and give voice to the universe's most
profound truths.

In this light it becomes clear that James's philosophical affiliations
were, in and of themselves, political statements. A vote for the unseen
world was a vote for a more meaningful life for those debilitated by ill
health. He contended that our epistemological and metaphysical positions inform and define how we value those members of society who are
afflicted by sickness and infirmity. James's politics of pathology went beyond challenging social perceptions, however. He also concerned himself with the practical consequences of medical materialism and specifically the therapeutic regimes under which mental and physical disorders
were treated. His activism in this area was to cost him dearly.

Therapeutic Campaigns

On March 17, 1898, the *Boston Medical and Surgical Journal* published a
vitriolic attack on "Professor James" and called for him to be disciplined
by Harvard College. The article described him as a "spokesman of me-
dievalism and an ally of quackery," someone who had "for a long time
allied himself with the enemies of scientific medicine." James's crime
was to have supported the cause of the local mind-cure community. His
sympathetic interest in "faith cure or mental science" was construed as
"another instance out of many of the 'highest culture,' so-called, returning to primitive barbarism." It was, the author explained contemptu-
ously, the intellectual equivalent of "the dog returning to his vomit."[95]

The article was written in response to a particular event, namely, a
hearing in front of the Committee on Public Health of the Massachusetts legislature, which had been held on the morning of March 2. It
concerned a bill that proposed that no person should be permitted to
practice medicine in Massachusetts who was not registered with the
Board of Registration in Medicine. Many, including James, considered
that such a bill would unfairly prohibit the work of the state's "unorthodox" healers. At the hearing James and other opponents of the bill were
introduced by Harrison Barrett, the editor of the spiritualist periodical
Banner of Light, who claimed to represent officially a range of healers,
many of whom typically advertised their services in the *Banner*. These
included "spiritualists, electricians, osteopathists, metaphysicians, magnetic healers, spiritual healers, botanic physicians and hydropathists."[96]

As James's son later attested, some of his father's medical colleagues never forgave him for allying himself with "the spokesmen of all the -isms and -opathies." Writing in 1920, ten years after his father's death, Henry James III remarked that "to this day references to [James's] 'appearance' at the State House in Boston are marked by partisanship rather than understanding."[97]

At the time James found the whole ordeal acutely difficult, describing it to a close friend and doctor, James Jackson Putnam, as one of the hardest things he had ever done. "I never did anything," he told him, "that required as much moral effort in my life." He went on to explain that he felt his vocation was to "treat of things in an allround manner" rather than make "*ex parte* pleas to influence (or seek to) a peculiar jur[y]."[98] In this instance, however, James had felt forced by the "medical brethren" into taking sides. He delivered his public protestations about the bill in the full knowledge of how his actions would be viewed by many colleagues at the Harvard Medical School. Despite this James felt inescapably driven by his own conscience to take a stand. He described his opposition to the bill as an act of "civic virtue" and wryly compared himself to the French novelist and journalist Émile Zola and army investigator Colonel Georges Picquart, who had spoken out to challenge the controversial imprisonment of the Jewish Captain Alfred Dreyfus.[99] James felt it his duty to maintain his stance, however much abuse it garnered him. Evidently, from his perspective, the stakes for society at large were extremely high.[100]

This was the second time that James had publicly defended the local lay healing community. The first time, four years earlier, had also been in response to an attempt to introduce legislation regulating all medical and surgical practices. His second protest, in particular, exhibited a distinctly anti–medical establishment polemic. His initial objections, which survive in the form of a letter written to the editor of the local *Transcript* newspaper, are relatively mild-mannered in their tone, but the address he delivered at the hearing of the medical registration bill in 1898 was far more confrontational. It would appear that, by that point, James had lost his patience with the medical profession. He referred to them, more than once on this occasion, as having adopted a "fiercely partisan attitude" akin to that of a "powerful trades union, demanding legislation against the competition of the 'scabs.'" In these sections of the address, he strongly implied that in bringing about the bill the "medical politicians" who ran the affairs of the Massachusetts Medical Society had been motivated, to some extent, by financial greed. In private he claimed that the only ostensible grounds for the bill that had been put forward were those of "personal dislike," and he compared the medical profession's actions to anti-Semitism.[101]

It would be wrong to assume that James's defense of the mental healers arose simply from an insistence on the spirit of fair play or a fondness for supporting the underdog. James's address contained an explicit challenge to medical authority. In one passage he explained to his audience that the medical profession should not wonder at the readiness of the public to seek out alternative healers given the sorry past of their own profession: "The history of medicine is a really hideous history," he exclaimed, "comparable only with that of priestcraft; ignorance clad in authority, and riding over men's bodies and souls."[102] He pointed out, moreover, that such prejudices are made worse, rather than better, when those who possess them "band themselves together in a corporate profession." As a medical student, he recalled, he was conscious that any student would have been "ashamed to be caught looking into a homeopathic book by a professor. We had to sneer at homeopathy by word of command."[103]

James's ire had been aroused, in part, by the orthodox physicians' dogmatic conception of what should constitute medicine and how disease should best be understood. James used his address to the Massachusetts Committee on Public Health, not just to remonstrate against the unfair treatment of the local, unlicensed healers but also, to strike a blow at the philosophical heart of contemporary medicine. At a time when the project of psychiatry had become synonymous with pessimistic notions of heredity and degenerate brains, James tried to carve out a place for psychogenic illness and the promise of nonphysical therapies. James's objection to the proposed legislation was simple. The medical profession was wrong to curtail and dismiss the practices of the lay healers, because they worked. Whether the doctors liked it or not the mind-curists and faith healers were producing results that were "patent and startling." James acknowledged that "hardly any medical subject [had] made greater progress in the past twenty years than that of neurology" but pointed out that the gains were "almost exclusively in the way of anatomy, symptoms, classification and diagnosis." Any drugs that had been found may have proved effective over "momentary states," but they had had little bearing on the patient's permanent cure.[104] Meanwhile, he explained:

> Of all the new agencies that our day has seen, there is but one that tends steadily to assume a more and more commanding importance, and that is the agency of the patient's mind itself. Whoever can pro-

duce effects there holds the key of the situation in a number of morbid conditions of which we do not yet know the extent, for systematic
experiments in this direction are in their merest infancy.[105]

James asserted in his address that the proposed medical registration
bill, by outlawing the mind-curers, would "kill [these] experiments out-
right." Under these circumstances the public health committee would be
converting "the laws of this commonwealth into obstacles to the acqui-
sition of truth."[106] For James the politics of law-making and the issue of
epistemological freedom were, on this occasion, one and the same. The
fate of medical knowledge, and the fate of the afflicted, lay in the committee's hands. He supported the right of the sufferer to conceive of their
symptoms in an alternative philosophical paradigm, outside of medical
materialism, not only because this might reconfigure the value of their
experience but also, because it might have pragmatic consequences for
their treatment and recovery.

As far as James was concerned the mixture of ignorance and authority
that characterized contemporary medicine was a dangerous brew and
nowhere more so than in the lunatic asylums. Aside from James's teaching interests in the field of psychiatry, and the class visits he arranged
to a variety of institutions for the insane, he was also closely involved
with such places for personal reasons.[107] His cousin Kitty Prince was
considered to suffer from some form of delusional insanity and spent
many years at the local Somerville asylum. After one of his regular and
distressing visits to see her, James wrote the following account to his
wife, Alice:

> Kitty has been badly excited & suicidal for a month past, now refuses
> food and is fed forcibly—looks hardly recognizable, poor soul. . . .
> They are pursuing the sodden Asylum routine of stomach pump etc.
> By Heaven I trust that the Doctors will catch it at the day of judge-
> ment for their besotted incompetence and authority combined. I gave
> Cowles a piece of my mind; but I fancy it made no practical impres-
> sion. The routine is for the sake of protecting themselves against fu-
> ture accusations of negligence. The best good of the patient is the last
> thing thought of.[108]

This passionate repudiation of asylum practices found an outlet, several
years later, when James became acquainted with Clifford Whittingham
Beers.

Beers was a thirty-year-old Yale graduate when he first contacted
James in 1906 and sent him a manuscript describing the several years

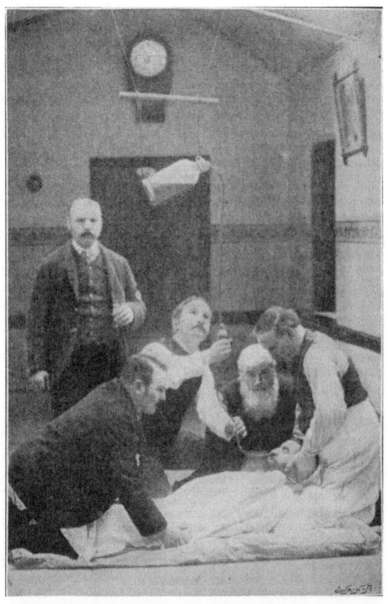

5.2 An image from an 1894 article in the *British Medical Journal* on "The Forcible Feeding of the Insane." (Wellcome Collection, Attribution 4.0 International [CC BY 4.0].) The author, an assistant medical officer at the North Wales Counties Asylum, recommended a particular configuration of stomach pump and other equipment. He stated that one advantage of feeding via the nose rather than the mouth was that "the patient seems to realise sooner that he is mastered." (Wm. W. Herbert, "The Forcible Feeding of the Insane," *British Medical Journal*, no. 1731 [March 1894]: 462.)

of his life that he had spent inside various psychiatric institutions. The manuscript, which was later published as a book entitled *A Mind That Found Itself: An Autobiography*, detailed Beers's descent into a state of severe mental distress as a young man, his suicide attempt and subsequent internment at both private and state lunatic asylums. His self-avowed intention in writing about these experiences was to "prick the civic conscience into a compassionate activity, and thus bring into a neglected field earnest men and women who should act as champions for those afflicted thousands least able to fight for themselves."[109] The "neglected field" to which he referred was, broadly speaking, the care of the insane. Beers's campaign was not targeted solely at improving conditions for those confined to the asylums, however, he also aimed at the prevention of mental disorder.

James's response to the manuscript was enthusiastic and encouraging. He urged Beers to finish his account and publish it and assured him: "You have no doubt put your finger on the weak spots of our treatment of the insane, and suggested the right line of remedy." He added that he had "long thought that if I were a millionaire, with money to leave for public purposes, I should endow 'Insanity' exclusively."[110] This letter marked the beginning of a collaboration that lasted until James's death, four years later. Over the course of these years the two men exchanged scores of letters, and James became intimately involved in Beers's developing campaign.

At a meeting in early 1907 Beers sought James's help and advice concerning the publication of his work. James gave him permission to use his original letter praising the manuscript as a preface to the book, authorized Beers to use his endorsement whenever it might be useful and offered to write a letter of introduction to one of his own publishers Longmans, Green. Beers's gratitude was exuberant. He told James that if he had had the "choice between the use of your name in the furtherance of my project, and the use of millions of dollars of one or a dozen of our men of vast wealth, I should have chosen the good name rather than the great riches." In April 1907, less than two weeks after receiving Beers's manuscript vouched for by the "magic name" of James, Longmans agreed to publish it.[111]

Publication of his own experiences comprised only one half of Beers's overall project. He also sought to establish an organization that could influence and oversee positive changes with regard to the treatment and prevention of insanity. His master plan was to establish a national society accompanied by a system of state societies, all of which would begin with the appointment of a board of honorary trustees. These men and women, and their patronage of the cause, would help him realize the necessary reforms. Here James was, once more, instrumental to Beers's campaign. He was the first person to agree to serve as an honorary trustee and went on to write letters urging several of his friends to do likewise. Those whose participation he secured included prominent figures, such as the banker and philanthropist Colonel Henry Lee Higginson and Harvard professor Francis G. Peabody. When the National Committee for Mental Hygiene was founded, two years later in 1909, James became a member of the executive committee.[112]

James also played an active role in seeking funds for Beers's initiative. He personally donated a thousand dollars of his own money and approached various associates for contributions and grants, including Charles William Eliot, the ex-president of Harvard, who was, by that point, a member of the General Education Board.[113] He also volunteered to write a letter of appeal to the wealthy philanthropist John D. Rockefeller (1839–1937), something he had not done for anyone else.[114] According to his wife, Alice, James was more interested in this movement "than in any other cause with which he is identified."[115]

Although funding from Rockefeller was not forthcoming at this stage in the campaign the latterly established Rockefeller Foundation made a series of significant contributions to the National Committee for Mental Hygiene. By 1920 the foundation had donated $289,000, or 38 percent of the total gifts received by the organization. Ultimately, Beers's campaign succeeded in initiating investigations into the conditions prevalent in the insane asylums, making them public knowledge and enlisting support to improve them. The committee also influenced medical school curricula and, in 1917, began producing its own journal, *Mental Hygiene*. Over time the committee came to focus more on programs dealing with the prevention, diagnosis, and treatment of mental disorder outside of the asylums, creating new training fellowships and becoming the chief agency for raising psychiatric awareness in the United States.[116] It is impossible to determine how instrumental James was to Beers's project and its eventual outcomes, but Beers perceived his support and involvement to have been crucial.

Beers's heartfelt appreciation of James, and all his campaigning efforts, is evident in a letter that he wrote to him in 1910. Concerned by a note of finality in James's last correspondence Beers feared his friend was critically ill and was moved to describe the depth of his affections for him.

> No matter how many friends come as rewards in my work, none can ever take your place. The circumstances which gave rise to our friendship cannot recur. Your opinion regarding the value of my book in

manuscript and my plans in general was the first supporting and authoritative opinion obtained. You had faith in me, and weren't afraid to say so. From that moment success was assured.

I doubt if you fully appreciate what your unfailing and generous support has meant to me. Many a time, when facing a difficult situation in the development of my plans, have I said, "Well, if so and so won't do it, Mr James *will*"—and you always did.[117]

For his part James deeply admired Beers and all that he had achieved. He replied that

it "sets me up" immensely to be treated by a practical man on practical grounds as you treat me. I inhabit such a realm of abstractions that I only get credit for what I do in that spectral empire; but you are not only a moral idealist and philanthropic enthusiast, (and good fellow!) but a *tip-top man of business* in addition; and to have actually done anything that the like of you can regard as having helped him is an unwonted ground with me for self-gratulation. . . . I esteem it an honor to have been in any degree associated with you.[118]

It is apparent that James's respect for Beers was inspired, in some measure, by his friend's practical efficacy. James tended to think of himself, in comparison, as "an impractical man," ill-versed in matters of an organizational nature.[119] But there is also evidence that James and Beers shared much common ground with respect to the conceptualization of mental disorders and how best they might be treated. At the beginning of his book *A Mind That Found Itself*, Beers insisted that what followed was in part biography, rather than autobiography, as "in telling the story of my life, I must relate the history of another self—a self which was dominant from my twenty-fourth to my twenty-sixth year." This section of his account might, he continued, be called the "history of a mental civil war, which I fought single-handed on a battle-field that lay within the compass of my skull."[120]

Beers's depiction, of two selves competing within one psyche, vividly resembled James's own psychopathological ideas about "the divided self," which were discussed earlier. James also maintained that the existence of secondary, subliminal strata of consciousness were likely to be the source of all sorts of "pathological cases," including "insane delusions" and "psychopathic obsessions."[121] In this way Beers's account of his own case provided corroboratory evidence for James's conviction that many forms of insanity were best understood according to a psychological, rather than physiological, etiology. In his book Beers addi-

5.3 The first page of the original draft of James's letter to Rockefeller, which was sent several months later in June 1909. The name of the original intended recipient "Phillips," which is crossed out, may have been a misspelling of Phipps. Henry Phipps was an American philanthropist whom Beers also wished to approach for funds at this stage. (Draft of William James to John Davison Rockefeller, January 31, 1909, William James Papers [MS Am 1092.9 (3553)], Houghton Library, Harvard University.)

tionally, like James, brought attention to the need for individualized forms of moral treatment as opposed to routine adherence to the medical textbooks. He implied in his autobiography that he might have recovered more quickly had his doctors taken the time to engage him in conversation and patiently reason with him in his deluded state, rather than force-feed him medicine that produced no positive effects. "The greater initial expenditure entailed in the individual treatment of a case would," Beers insisted, "be offset by the more rapid recovery and earlier discharge of the patient."[122] In one telling incident James accidentally addressed Beers, on the envelope of a letter, as "Dr." Inside he wrote, "I have inadvertently called you Dr. on the envelope—let it stand! you are better than a Doctor."[123]

The contents of the written appeal for funds that James drafted for Rockefeller's attention summed up his and Beers's shared conception of the campaign to which they both committed a considerable amount of time and money. James wrote:

> During my life as a "psychologist" ~~In my earlier years, and later too,~~ I have had much to do with ~~lunatic~~ our Asylums, and I have had so painfully borne in upon me the ~~total mass~~ massiveness of human evil which the term "insanity" covers . . . not only ~~do~~ are sodden routine and ignorance ~~prevail there in them~~ too often found [in our asylums] but ~~the~~ public opinion needs educating in the matter. . . . Everyone is liable to [insanity] if the strain-point is overpast. . . . What should be regarded as a common ~~cerebral~~ functional disease ~~gets~~ is treated as a social stigma. . . . I end by merely ~~urging you~~ begging you and urging you to give to this project your very best attention. *Nothing I know of is more important.*[124]

This entreaty contained many of the elements of James's politics of pathology. It outlined his concerns that those labeled "insane" suffered prejudice and mistreatment and disputed the supposed gulf between the sane and the mad. It also alluded to his preferred approach to insanity as a "functional disease," a *psycho*pathological state that is amenable to mental healing methods, rather than an incurable and hereditary brain disease. Finally, it stated clearly that this cause, above all others, mattered intensely to James in the last years of his life.

Afterlife

Fit to Live

In 1904 James completed a questionnaire composed by one of his graduate students, James Bissett Pratt (1875–1944), who was interested in studying individual experiences of religion. James was asked, among other things, whether he believed in "personal immortality." He responded, "Never keenly; but more strongly as I grow older." In answer to the second half of the question, "If so, why?" he added, "Because I'm just getting fit to live."[1] By 1904 James was sixty-two years old, had been suffering from a debilitating heart complaint for six years, and had endured episodes of severe "nervous prostration" for longer still. Regarding his physical health he had, arguably, never been in worse shape. By this stage in his life, however, James's sense of his own self-worth and his faith in his ability to make a meaningful contribution to society, were no longer chained to his assessment of his bodily condition. His response to Pratt's question epitomizes the central narrative of this book, which may be read as a commentary on James's changing understanding of what it means to be "fit to live."

James's deliberations on this theme date from his late twenties when he began debating with himself and various correspondents about whether he should take his own life. His reading lists, private notes, and letters from this period testify to the persistence with which he pursued these suicidal musings, which circled around his ill health and its consequences for himself, those close to him, and society at large. He was acutely aware of the impact of his invalid status on his family and friends and asked himself: "Shd. sympathy go so far as to dictate suicide? As when I sick, become but an eyesore & stumbling block to others."[2]

The experience of pain loomed large in James's reasonings and took on an ethical and a metaphysical significance. Within his worldview, illness and pain were evil incarnate, and to have them dwell indefinitely inside of him was a difficult cross to bear. A palpable air of shame hung around James's private wranglings. He confessed that he clung to "the idol of the unspeckled," and the thought of evil taking "permanent body" *in* him was enough to make him "give up all."[3] This ideal of the untainted, the idolatry of health, was a powerful and growing presence in late nineteenth-century America, and James was far from alone in espousing it. In the context of his own infirmities, however, it constituted a commanding rebuttal of his right to exist, and there is evidence that this sense of shame never truly left him. After his death, his friend and Harvard colleague, philosopher Josiah Royce (1855–1916), lamented the irony of his own belief that James had "shortened his life by fussing over his heart—not that he wanted to live, but that he thought it somehow a disgrace to have diseases and was always trying to cure himself. . . . If he had only let it alone and thought of something else."[4]

James's inability to "let it alone" was borne of his abiding commitment to the hygienist conviction that the preservation of one's personal health was a civic duty. In line with the widely influential writings of Herbert Spencer, who described offenses against the laws of health as "*physical sins*," James referred to himself, in the context of his invalidism, as a "sinner."[5] Similarly, they both described the "evil consequences" of such "bodily transgressions" for future generations as a social "crime." Despite James's documented early opposition to many of Spencer's theories, his medico-ethical values and his environmental concept of health continued to play a significant role in James's thinking throughout much of his life.

Within Spencer's philosophy, the natural world was elevated to the role of a deity; the environment constituted the passive fashioner of humankind. The race must, in Spencerian terms, "correspond" or suffer, and only the fittest would survive. He identified perfect physiological life with perfect adaptation, and it was this definition of health that James appears to have taken to heart. In the early 1890s, for example, when he composed and first delivered his *Talks to Teachers on Psychology*, his lectures and notes indicate that he commended Spencer for this insight. James agreed with him that "adaption to the environment is the watch word" and that an educated citizen may be defined as one whose conduct "shall fit him to his social and physical world."[6] While James acknowledged that worthwhile human activity extends beyond this "biological conception" of existence, he maintained that survival and health must, out of necessity, come first.

In parallel with this fidelity to the hygienic ideal James also maintained, however, a long-standing commitment to the value of the invalid and their potential contribution to the moral wealth of society. As a response, in part, to the increasing intensification of the degeneration debates, James developed these private convictions into what I have termed a politics of pathology. Across a number of public addresses from the mid-1890s onward, he publicly challenged the opprobrium associated with sickness and infirmity and promoted in its stead a series of laudable social roles for "the degenerate" and the invalid. These were constructed around the themes of genius, insight, heroic resilience in the face of suffering, and the potential metaphysical significance of these acts of endurance for humankind's ultimate salvation.

When James delivered his lectures on *The Varieties of Religious Experience* at the turn of the twentieth century, he went further still. In his *Talks to Teachers*, biological prowess was positioned as the necessary foundation for all other human endeavors, including the pursuit of "ethical utopias" and "insights into eternal truth," but in *The Varieties*, physical strength was allied with aggression and social disruption, and the epistemological experience of illness was exalted. James distinguished the privileged vantage point of those who have encountered the darkest depths of painful evil at firsthand and recognized their contributions to philosophical wisdom. Only those who are "sick souls" have access to the profoundest truth, he declared. In this way James had, by the time he came to answer Pratt's questionnaire, become the outspoken champion for an avowedly anti-Spencerian understanding of who is best fitted to life. Throughout *The Varieties*, however, the environment, albeit in a pluralistic sense, remained in a position of authority within James's thought.

In these lectures James challenged the opinion that there can ever be "one intrinsically ideal type of human character" determined by the evolutionist's criteria. "How is success to be absolutely measured," he asked, "when there are so many environments and so many ways of looking at the adaptation?" Rather than the biological world, for example, James asked his audience to consider "the larger environment of history" or the metaphysical environment of the "unseen world." He concluded that human achievement "cannot be measured absolutely; the verdict will vary according to the point of view adopted."[7] But although James contested the preeminence of the natural environment, he did not question its sovereignty over the realm of physiological activity. He undermined the normative status of biological supremacy but not its conceptual definition.

It was not until 1906, in his article on Giovanni Papini and his lecture on "The Energies of Men," that James abandoned the environmental formula altogether, even as a measure of "physical well-being." Inspired by Papini's pragmatic philosophy, and the personal testimony of his friend Wincenty Lutoslawski about his miraculous yoga cure, James proposed that humanity's potential, physical and mental, be left open and unrestricted. Rather than determining the nature of the world in which we live and then adjusting ourselves accordingly, James proclaimed that we should first seek to define the furthest reaches of our own creative human powers. He replaced the expectation that the human race need mold itself to the contingencies of its environment with the idea of "Man-Gods" who strive to create the kind of reality they desire.

This boldly humanist perspective only became viable to James in the context of his later theories about medical truth and those who claim to have discovered it. For James, the success of professional and lay mental healing techniques that treated the individual at the level of their personality, rather than deconstructing them into general, physiological processes, was instrumental in his own willingness to concede intellectual authority to the kinds of insights that he labeled collectively as "religious." From his perspective the defining characteristic of religious or moral truths was their claim that "events may happen for the sake of . . . someone's personal significance," as opposed to the conventional scientific assumption that the world is, in its "innermost nature[,] . . . a strictly impersonal world."[8]

From James's standpoint, the human-centered vision of reality theorized by Papini, and exemplified by Lutoslawski, comprised the ultimate liberation from the medical materialist laws of life and health; the laws that had, for many years, informed James's perception of himself as an invalid and humankind's capacity for regeneration. It vanquished the last vestiges of the notion that any aspect of an individual's fitness inheres in a biological matrix of body and environment. Instead, James concluded, the "unexplored powers and relations of man, both physical and mental, are certainly enormous; why should we impose limits on them *a priori*?"[9]

Moral Medicine

James interrogated contemporary concepts of health and disease and their implications for the individual and society, and, in doing so, he laid bare the limitations and value judgments at the heart of the modern medical project. He drew attention to the culturally contingent nature of medical labels, especially within the psychiatric literature, and he explicitly challenged the conceit of the "so-called 'normal man.'"[10] He did not oppose diagnosis per se, however. James used and even on occasion applauded the novel diagnostic categories deployed by his medical associates; but he handled them in a distinctly pragmatic manner. He eschewed the tendency of many of his contemporaries to affiliate diagnosis with discrimination, instead stressing the nearness of "normal" and "abnormal" behaviors. Diagnostic indications were conceptual stepping-stones to useful insight and ameliorative action, not social stigmata used to identify, demean, and disempower particular groups.

James's relationship with the concept of "health" was equally nuanced. Despite challenging many of the assumptions and definitions assigned to it, James continued to promote it for himself and others. Indeed, he remained firmly wedded, throughout his life, to the ethical conviction that "*health* is the only good."[11] If James came to reject popular interpretations of health, what personal meaning did it hold for him? A posthumous account is suggestive. His friend and colleague James Jackson Putnam wrote that although James had been "much of an invalid," he believed that "one should industriously cultivate the bearing, the expression, and the sentiments that go with health, and one of his former pupils has recently told me of his making an appeal to his college class on this subject."[12]

This testimony is notably redolent of the mind-curists' "healthy-minded" approach to life and also of James's own writings on emotion. In his address on "The Gospel of Relaxation," he emphasized the hygienic potential of fostering particular emotions at will, via the deliberate decision to act as if you are already feeling them: "action and feeling go together; and by regulating the action, which is under the more direct control of the will, we can indirectly regulate the feeling, which is not."[13] I propose that James comprehended health, primarily, not as an objectively measurable set of criteria but as a subjective state of feeling: an emotion of health.

This hypothesis is consistent with the most fulsome description of "health" that James committed to paper. According to his notes, he ended his 1895 lecture on "The Effects of Alcohol" with a quote from the writer Charles Leland that he remembered having read for the first time thirty-three years earlier. This passage, which had clearly made a lasting impression on him, defined the "ideal" of health as "the most desirable state of mind." James repeated Leland's insistence that "simple, sober, perfect *health* on a fine Indian-summer morning" is a feeling of "such exhilaration"; the excitement of his "own bounding life-blood" was the "best excitement" that he had ever known.[14]

This effervescent depiction also coheres with James's enthusiasm for two particular medical frameworks, Beard's neurasthenia and Janet's psychasthenia, which both established some form of energy as the currency of health. In his "Energies of Men" essay, inspired in part by Janet's research, James focused on the medical significance of "the amount of energy available for running one's mental and moral operations by." James explicitly acknowledged the subjective nature of this energy level, conceding that although "everyone knows in his own person the difference between the days when the tide of this energy is high in him and those when it is low ... no one knows exactly what reality the term energy covers when used here, or what its tides, tensions and levels are in themselves." Despite the unquantifiable nature of this energy, however, he stated that "to have its level raised is the most important thing that can happen to a man." I argue that this "feeling," the "phenomenon of feeling more or less alive," was the consistent kernel of James's definition of health throughout his life.[15]

This definition also makes it possible to reconcile James's insistent glorifying of health, and the significant degree of personal responsibility that he attached to it, with his equally vocal politics of pathology and his valorization of illness and suffering. He appears to have advocated for both health and unhealth. From James's later-years perspective, however, health and illness were not opposites, nor did they exist on some kind of continuum or spectrum with serious illness at one end and perfect health at the other. Instead, these two categories of experience could happily coexist in one person. This position is exemplified in the case study of a friend, which he cited in "The Energies of Men," whom he described as "the most genuinely saintly person I have ever known." Despite suffering from cancer of the breast, James testified that her mindcure principles had "kept her a practically well woman for months after she should have given up and gone to bed. They have annulled all pain and weakness and given her a cheerful active life, unusually beneficent to others to whom she has afforded help."[16] This spirit of endurance, this "firmer, more elastic moral tone," as James phrased it, represented, for him, the essence of health. The ultimate good was not freedom from disease but the sentiment of irrepressible vitality in the presence of the evils of illness, pain, and death.

According to James the mind-cure or mental healing movement, to which his friend belonged, constituted a "wave of religious activity," and this description spoke to the central purpose that he conferred on religious and ethical belief systems.[17] He persistently extolled their function as stimuli; supreme initiators of human energies; faiths that brought forth the desirable "moral tone" of health. In his texts on religion, there is a sense in which James adopted the air of a physician dispensing meta-physical prescriptions. In *The Varieties of Religious Experience* and *Pragmatism*, he proposed different ethico-religious solutions to suit the differing temperaments of his patients, but, for him, religion and health were always intimately connected.

James conceived of medicine as a radically *moral* endeavor, meaning both one imbued with ethical judgments, but also in the broader nineteenth-century sense of relating to the domain of psychology and specifically emotions and the will. In other words, James affirmed the ethical content of medical ideas and also the medical import of "moral" matters. Many of his texts, including those on habits, emotions, pedagogy, ethics, and religion, that haven't traditionally been linked with the pursuit of health, assume a hygienic or healing aspect when they are situated in the context of James's interests and the cultural concerns of his era. James came to believe that medical knowledge existed outside of the confines of physiology and that "all kinds of health, bodily mental and moral are essentially the same, so that one can go at them from any point."[18] In his obituary of James, his friend Putnam recalled "the earnestness with which he said to me, two years ago, that the results he had achieved were, in kind, just those he had aspired to achieve." Those aspirations were, namely, to accentuate "certain tendencies in the minds of thinking men which he believed to be wholesome and of vital significance."[19]

His audience undoubtedly discerned a therapeutic quality to his work. On one occasion James wrote to his wife from his travels that he had met a woman who "said she had my portrait in her bedroom with the words written under it 'I want to bring a balm to human lives'!!!!! Supposed to be a quotation from me!!!"[20] On a more serious note, an acquaintance confided in him that his essay on suicide, "Is Life Worth Living?," and *The Varieties of Religious Experience* had saved her life.[21] Similar tributes may be found in the letters sent to James's son, Henry James III, after his father's death. One such correspondent acknowledged that his personal obligation to James "in the way of moral encouragement in the face of difficulties which seemed to have no possible solution" was "greater than to any one else, and I cherish his memory as a personal deliverer."[22] One explained that James's books had "transformed all life" for her and assured Henry that "every mail must bring you grists of such sentiments for thousands of lives, like mine, were helped towards sanity and usefulness by your father."[23] Another stated succinctly, "The world mourns your father, as Philosopher, Teacher, Uplifter of men."[24] James's readers felt the compelling emotional intensity of his relationship with the world of ideas, and they shared his belief that certain philosophies can offer profound comfort, inspiration, and healing to those who suffer. A dedicated if unconventional physician, James's writings were not just words on a page, they were regenerative doses of moral medication.

ACKNOWLEDGMENTS

My interest in William James began when I wrote an essay on James and Stoicism during my MA at the Wellcome Trust Centre for the History of Medicine at University College London. The research for that essay grew, eventually, over many years, into this book. My work on James throughout this time has owed an inestimable amount to Sonu Shamdasani. I have valued both his teaching and support and his inspiring commitment to the historical and creative potential of biographical approaches. Sonu also introduced me to James scholars Jeremy Carrette, Ramón del Castillo, Francis Neary, and Eugene Taylor, whose informal conversations and published contributions to the field have all played a significant role in the development of my thinking.

During my time at UCL, I was very grateful to a number of colleagues from the history of medicine and psychology, in the UK and further afield, who read my work and gave me feedback. These included Michael Neve, Robert Kugelmann, Stephen Jacyna, and, most notably, Andrew Wear, who gave up his time to comment on an early draft of this book. I was also lucky to be part of an immensely stimulating and collaborative academic community. Thank you Corina Doboş, Sally Frampton, Sheldon Gosline, Matei Iagher, Laura Ishigaru, Tom Quick, Sarah Marks, Dee McQuillan, Steve Ridge, Andreas Sommer, Erin Sullivan and especially Jane Berridge and Sarah Desmarais

My James research would never have come to fruition in this form, however, if it wasn't for the collegial home I found at the Centre for the History of the Emotions, Queen Mary University of London. My post there provided both the time I needed and a rich intellectual environment in which to develop my ideas into a book. My knowledgeable and supportive colleagues included Elena Carrera, Jules Evans, Rhodri Hayward, Andrew Mendelsohn, and Tiffany Watt-Smith, and I am particularly indebted to Sarah Chaney, Thomas Dixon, and Helen Stark for their insightful comments on the draft proposal and chapters for this book. Thomas's expert mentoring was instrumental to the completion of my manuscript and hugely appreciated during the challenges of the COVID lockdowns.

Thank you to Karen Darling at the University of Chicago Press for her encouraging belief in this project at every stage. The two reader's reports secured by the Press were characterized by a spirit of generosity and helpfulness, and I am extremely grateful to their anonymous authors. Thanks also to Fabiola Enríquez Flores for her very friendly and efficient assistance and to Anne Strother and Mark Reschke for their expertise and input.

While working on this book I have benefited enormously from the support, advice, and expertise of friends and associates outside of academia. Thank you especially to Louise Popple, Sivan Halevi, and all those with whom I worked at the BBC. I draw on the skills and experiences I gained during my time as a filmmaker every day in my research and writing and would particularly like to thank Doug Carnegie, Andrew Cohen, and Caroline van den Brul.

In the course of my research, I have been assisted by library staff and archivists at the Houghton Library, the Countway Library of Medicine, the Wellcome Library, Senate House Library, and UCL Library. Their help and knowledge has been much appreciated, and I would especially like to mention Zahra Garrett at Harvard and the team at the Wellcome Library for their unwavering enthusiasm and kindness. On the topic of source materials, I would also like to extend my heartfelt thanks to James scholar James Anderson, who went to some lengths in order to help me obtain a particular set of archival papers.

Without the generous funding provided by the Wellcome Trust, the research for this book would not have been possible. In addition, the Royal Historical Society and UCL's graduate school both made significant contributions to the costs of my research trips and travel to overseas conferences. The inclusion of several of the images in this book was made possible by an Isobel Thornley grant administered by the University of London.

Material previously printed in Emma Sutton, "Interpreting 'Mind-Cure': William James and the 'Chief Task . . . of the Science of Human Nature,'" *Journal of the History of the Behavioral Sciences* 48, no. 2 (2012), has been reused in chapter 3 of this book with the kind permission of the publisher, and in chapter 1, I draw on ideas and sources originally discussed in Emma K. Sutton, "When Misery and Metaphysics Collide: William James on 'the Problem of Evil,'" *Medical History* 55, no. 3 (2011).

Finally, I would like to thank my parents, Helen and Michael Sutton; my brother, Timothy Sutton; my husband, Richard Penn; and our sons, Albert and Josiah. Their love and loyalty are an essential part of everything I do.

NOTES

Introduction

1. Ignas K. Skrupskelis, "Introduction," in *The Works of William James: Manuscript Lectures*, ed. Frederick H. Burkhardt, Fredson Bowers, and Ignas K. Skrupskelis (Cambridge, MA: Harvard University Press, 1988), xxvii, xlvii, xlviii, li. Francesca Bordogna makes the case that James's tendency to trespass across disciplinary boundaries was, for him, an end in itself and reflected his desire for a new social configuration of knowledge. Francesca Bordogna, *William James at the Boundaries: Philosophy, Science and the Geography of Knowledge* (Chicago: University of Chicago Press, 2008).

2. William James, *The Principles of Psychology*, vols. 1 and 2, ed. Frederick H. Burkhardt, Fredson Bowers, and Ignas K. Skrupskelis, The Works of William James (Cambridge, MA: Harvard University Press, 1981); William James, *The Will to Believe and Other Essays in Popular Philosophy*, ed. Frederick H. Burkhardt, Fredson Bowers, and Ignas K. Skrupskelis, The Works of William James (Cambridge, MA: Harvard University Press, 1979); William James, *The Varieties of Religious Experience*, ed. Frederick H. Burkhardt, Fredson Bowers, and Ignas K. Skrupskelis, The Works of William James (Cambridge, MA: Harvard University Press, 1985); William James, *Pragmatism*, ed. Frederick H. Burkhardt, Fredson Bowers, and Ignas K. Skrupskelis, The Works of William James (Cambridge, MA: Harvard University Press, 1975).

3. William James to Hugo Münsterberg, September 2, 1896, Ignas K. Skrupskelis and Elizabeth M. Berkeley, eds., *The Correspondence of William James*, vol. 8 (Charlottesville: University Press of Virginia, 2000), 197; "William James Dies; Great Psychologist," *New York Times*, August 27, 1910; "Notes," *Philosophical Review* 19, no. 6 (1910): 694; John Jay Chapman, "William James: A Portrait," *Harvard Graduates' Magazine* 1910, 238.

4. The medical aspects of James's thought and its development have received little sustained attention to date. I discuss James's conceptualization of evil and its relationship with illness in Emma K. Sutton, "When Misery and Metaphysics Collide: William James on 'the Problem of Evil,'" *Medical History* 55, no. 3 (2011), and I give a detailed account of his personal and professional dealings with the American mind-cure movement and the consequences for his epistemological and metaphysical ideas in Emma Sutton, "Interpreting 'Mind-Cure': William James and the 'Chief Task . . . of the Science of Human Nature,'" *Journal of the History of the Behavioral Sciences* 48, no. 2 (2012). James's focus on these and the other medical themes explored in this text may be found in Emma K.

Sutton, "Re-writing 'the Laws of Health': William James on the Philosophy and Politics of Disease in Nineteenth-Century America" (PhD thesis, University College London, 2013). Eugene Taylor reconstructs James's research into nineteenth-century theories about the subconscious self at length and explores some of the implications for James's interest in psychotherapeutics: Eugene Taylor, *William James on Exceptional Mental States: The 1896 Lowell Lectures* (New York: Charles Scribner's Sons, 1982); Eugene Taylor, *William James on Consciousness beyond the Margin* (Princeton, NJ: Princeton University Press, 1996). Relevant short contributions include Ralph Barton Perry's and Daniel Bjork's discussions of James's interest in psychopathology and mental therapeutics in Ralph Barton Perry, *The Thought and Character of William James: As Revealed in Unpublished Correspondence and Notes, Together with His Published Writings*, vol. 2, *Philosophy and Psychology* (London: Oxford University Press, 1935), 4–5, 91, 120–23, 156–60, 168–70, 318, 674; Daniel W. Bjork, *William James: The Centre of His Vision* (New York: Columbia University Press, 1988), 200, 208–14. Donald Duclow explores James's critique of the positive-thinking stance of the mind-curers as the basis for an assessment of the contemporary value of such approaches to illness in Donald F. Duclow, "William James, Mind-Cure and the Religion of Healthy-Mindedness," *Journal of Religion and Health* 41, no. 1 (2002). J. T. Matteson surveys James's defense of the mental healers in the face of the proposed state regulation of unorthodox medical practitioners: John T. Matteson, "'Their Facts Are Patent and Startling': WJ and Mental Healing," *Streams of William James* 4, no. 1 (2002); no. 2 (2002); Tad Ruetenik relates James's concept of "medical materialism" to the writings of Sigmund Freud and Jean-Paul Sartre: Tadd Ruetenik, "Fruits of Health; Roots of Despair: William James, Medical Materialism and the Evaluation of Religious Experience," *Journal of Religion and Health* 45, no. 3 (2006); Martin Halliwell discusses James's interests in psychic experimentation and therapeutic techniques in Martin Halliwell, "Morbid and Positive Thinking: William James, Psychology and Illness," in *William James and the Transatlantic Conversation: Pragmatism, Pluralism, and Philosophy of Religion*, ed. Martin Halliwell and Joel D. S. Rasmussen (Oxford: Oxford University Press, 2014); Paul Croce recounts James's attitude toward unorthodox medicine in chapter 2 of Paul J. Croce, *Young William James Thinking* (Baltimore: John Hopkins University Press, 2018), and Paul J. Croce, "James and Medicine: Reckoning with Experience," in *The Oxford Handbook of William James*, ed. Alexander Klein, accessed May 10, 2021, https://www.oxfordhandbooks.com/view/10.1093/oxfordhb/978 0199395699.001.0001/oxfordhb-9780199395699-e-29.

5. Henry Ingersoll Bowditch, *Public Hygiene in America: Being the Centennial Discourse Delivered before the International Medical Congress, Philadelphia, September, 1876* (Boston: Little, Brown and Company, 1877); Miriam Small, *Oliver Wendell Holmes* (New York: Twayne Publishers Inc., 1963), 55; William James, "Rationality, Activity and Faith," *Princeton Review* 2 (1882).

6. "Wilkinson, James John Garth," in *Dictionary of National Biography, 1885–1900* (London: Smith, Elder, & Co., 1885–1900), 61:271–27.

7. Accounts of James's interest in psychical research include Taylor, *Consciousness*; Krister Dylan Knapp, *William James: Psychical Research and the Challenge of Modernity* (Chapel Hill: University of North Carolina Press, 2017); Andreas Sommer, "James and Psychical Research in Context," in *Oxford Handbook*, ed. Klein, accessed May 10, 2021, https://www.oxfordhandbooks.com/view/10.1093/oxfordhb/9780199395699.001 .0001/oxfordhb-9780199395699-e-37.

8. The late nineteenth-century emphasis on health and personal responsibility, and James's attitude toward them, are explored in chapter 2, "Health and Hygiene."

9. Lucy Bending discusses nineteenth-century Christian understandings of physical suffering in the first chapter of Lucy Bending, *The Representation of Bodily Pain in Late Nineteenth-Century English Culture* (Oxford: Oxford University Press, 2000), and Naomi Strachan Donnelley explores the implications of the advent of anesthesia for religious and medical authority in America in her thesis: Naomi Strachan Donnelley, "Power and Mysticism in the Introduction of Anesthesia in Nineteenth Century America" (MD thesis, Yale University, 1998).

10. For James's account and critique of this economic evaluation of health and humanity, see William James, "Fifth Annual Report of the State Board of Health of Massachusetts (1874)," in *The Works of William James: Essays, Comments, and Reviews*, ed. Frederick H. Burkhardt, Fredson Bowers, and Ignas K. Skrupskelis (Cambridge, MA: Harvard University Press, 1987).

11. The historical use and conceptualization of the term "normal" is explored from a range of perspectives in Waltraud Ernst, ed., *Histories of the Normal and the Abnormal: Social and Cultural Histories of Norms and Normativity* (London and New York: Routledge, 2006).

12. I explore James's earlier attitude toward the idea of race betterment in chapter 2, "Health and Hygiene," and his later analysis of the degeneration literature in chapter 5, "Politics and Pathology."

13. William James, "*Degeneration*, by Max Nordau (1895)," in *Essays, Comments, and Reviews*, ed. Burkhardt, Bowers, and Skrupskelis, 513.

14. James's first biographer Ralph Barton Perry included an account of a period in James's early adulthood that he referred to as his "spiritual crisis." A considerable literature has grown up around this episode, which has been surveyed by Francis Neary. See Francis Neary, "Interpreting Abnormal Psychology in the Late Nineteenth Century: William James's Spiritual Crisis," in *Histories of the Normal*, ed. Ernst.

15. Ralph Barton Perry, *The Thought and Character of William James: As Revealed in Unpublished Correspondence and Notes, Together with His Published Writings*, vol. 1, *Inheritance and Vocation* (London: Oxford University Press, 1935), 79–80; Howard M. Feinstein, "The Use and Abuse of Illness in the James Family Circle: A View of Neurasthenia as a Social Phenomenon," in *Our Selves/Our Past: Psychological Approaches to American History*, ed. Robert J. Brugger (Baltimore: John Hopkins University Press, 1981); Gay Wilson Allen, *William James: A Biography* (New York: Viking Press, 1967), 149; Linda Simon, *Genuine Reality: A Life of William James* (New York: Harcourt, Brace and Company, 1998), 115; Howard M. Feinstein, *Becoming William James* (Ithaca, NY: Cornell University Press, 1999), 12–13.

16. Within the Jamesian literature, Louis Menand opens his book on the intellectual development of James and his confreres with a discussion of the American Civil War: Louis Menand, *The Metaphysical Club* (London: Harper Collins Publishers, 2001). George Cotkin stresses the imperialist tendencies of America's foreign policy as a crucial context within which to locate James's writings: George Cotkin, *William James: Public Philosopher* (Baltimore: John Hopkins University Press, 1990); James Livingston argues that James's pragmatism "functions as a narrative of the transition from proprietary to corporate capitalism." See James Livingston, *Pragmatism and the Political Economy of Cultural Revolution, 1850—1940*, Cultural Studies of the United States (Chapel Hill: University of North Carolina Press, 1994), xix. In terms of landmark texts and figures, Charles Darwin's writings have received a substantial degree of attention. Examples include but are not limited to Robert J. Richards, "James's Uses of Darwinian Theory," *A William James Renaissance: Four Essays by Young Scholars, Harvard Library Bulletin* 30,

no. 4 (1982); Paul Jerome Croce, *Science and Religion in the Era of William James*, vol. 1, *Eclipse of Certainty, 1820–1880* (Chapel Hill and London: University of North Carolina Press, 1995); Lucas McGranahan, *Darwinism and Pragmatism: William James on Evolution and Self-Transformation* (London: Routledge, 2019).

17. It has only been feasible to work across the entire Jamesian correspondence in this way thanks to the scholarship of Ignas Skrupskelis and Elizabeth Berkeley, who edited the twelve-volume collection of James's published correspondence. Ignas K. Skrupskelis and Elizabeth M. Berkeley, eds., *The Correspondence of William James*, 12 vols. (Charlottesville: University of Virginia Press, 1992–2004).

18. See, for example, William James to Henry James, October 25 [1869], Ignas K. Skrupskelis and Elizabeth M. Berkeley, eds., *The Correspondence of William James*, vol. 1 (Charlottesville: University Press of Virginia, 1992), 113–14.

19. Jeremy Carrette explores the theoretical conceptualization of the body, and in particular the sexual body, in relation to James's writings on religion in chapter 4 of Jeremy Carrette, *William James's Hidden Religious Imagination: A Universe of Relations* (London and New York: Routledge, 2013). In their introduction to a collection of essays entitled *Science Incarnate*, Christopher Lawrence and Steven Shapin allege that, within our contemporary culture, bringing "body and knowledge into contact . . . is occasionally taken as funny, sometimes as enraging, more often as just pointless." The chapters that comprise the main section of their volume address a variety of questions relating to the embodiment of knowledge as an academic topic. Steven Shapin and Christopher Lawrence, "Introduction: The Body of Knowledge," in *Science Incarnate: Historical Embodiments of Natural Knowledge*, ed. Christopher Lawrence and Steven Shapin (Chicago: University of Chicago Press, 1998). My study speaks to theirs and Carrette's in that I maintain that the reason James's ill health has not, to date, been taken more seriously in relation to his intellectual output is the contemporary reluctance toward making the kinds of general connection between body and knowledge to which they draw attention.

20. Perry, *Thought and Character*, 1:466. Perry distinguished between James's ongoing ill health, which he characterized as "Neurasthenia," and a limited period of "soul sickness" that culminated in a spiritual crisis. He implied that the former was an essential feature of James's physiological makeup and the latter a philosophical rite of passage.

21. See, in particular, the chapter on "Pragmatism and Humanism" in James, *Pragmatism*.

22. James only defined this term indirectly in his writings, but his usage of it invoked notions of biological determinism and contemporary depictions of humans as "conscious automata" that James took issue with. See chapter 1, "Misery and Metaphysics."

23. For instructive analyses of the significance of James's early adulthood to his later philosophy, see Perry, *Thought and Character*, vol. 1; Croce, *Science and Religion*; Taylor, *Consciousness*; Cotkin, *William James*; Croce, *Young William James Thinking*.

24. The Harvard University Press edition of the collected *Works of William James* edited by Frederick H. Burkhardt, Fredson Bowers, and Ignas K. Skrupskelis has been indispensable to my project in terms of finding, accessing, and contextualizing many of these smaller texts and manuscript sources.

25. William James, "Chauncey Wright," in *The Evolutionary Philosophy of Chauncey Wright*, vol. 3, *Influence and Legacy*, ed. Frank X. Ryan (Bristol: Thoemmes Press, 2000), 2.

26. William James to Thomas Davidson, January 8, 1882, Ignas K. Skrupskelis and Elizabeth M. Berkeley, eds., *The Correspondence of William James*, vol. 5 (Charlottesville: University of Virginia Press, 1997), 194. See, most notably, William James, "The Sentiment of Rationality," in *The Will to Believe*, ed. Burkhardt, Bowers, and Skrupskelis.

27. George Cotkin argues that James's writings were intended to function as public facing social commentaries in Cotkin, *William James*.

28. William James to Alice James, March 16, 1868; William James to Oliver Wendell Holmes, September 17, 1867, Ignas K. Skrupskelis and Elizabeth M. Berkeley, eds., *The Correspondence of William James*, vol. 4 (Charlottesville: University Press of Virginia, 1995), 200, 265; William James to Henry James, May 25, 1873, Skrupskelis and Berkeley, eds., *Correspondence*, vol. 1, 208.

29. William James to Sarah Wyman Whitman, September 18, 1902, Ignas K. Skrupskelis and Elizabeth M. Berkeley, eds., *The Correspondence of William James*, vol. 10 (Charlottesville: University of Virginia Press, 2002), 128.

Chapter One

1. Martin S. Pernick, "The Calculus of Suffering in Nineteenth-Century Surgery," *Hastings Center Report* 13, no. 2 (1983): 26. On the history of pain, see Joanna Bourke, *The Story of Pain: From Prayer to Painkillers* (Oxford: Oxford University Press, 2014); Rob Boddice, ed., *Pain and Emotion in Modern History* (Basingstoke: Palgrave Macmillan, 2014).

2. See, for example, Elizabeth B. Clark, "'The Sacred Rights of the Weak': Pain, Sympathy, and the Culture of Individual Rights in Antebellum America," *Journal of American History* 82, no. 2 (1995): 463–93; Michael Ruse and Robert J. Richards, eds., *The Cambridge Handbook of Evolutionary Ethics* (Cambridge: Cambridge University Press, 2017); Pernick, "Calculus of Suffering."

3. See, for example, Maureen Moran, "Hopkins and Victorian Responses to Suffering," *Revue LISA/LISA e-journal* 7, no. 3 (2009): 570–81; Strachan Donnelley, "Power and Mysticism"; Clark, "Sacred Rights of the Weak," 473.

4. Mary Robertson Walsh James to William James, November 21 [1867]; William James to Thomas Wren Ward, September 10, 1867, Skrupskelis and Berkeley, eds., *Correspondence*, 4:197, 231.

5. William James to Alice James, August 6, 1867; William James to Henry James Sr., September 5, 1867; William James to Henry James Sr., December 26, 1867; William James to Alice James, March 16, 1868 ibid., 187–89, 194, 243, 265.

6. William James to Henry James, December 26, 1867, Ignas K. Skrupskelis and Elizabeth M. Berkeley, eds., *The Correspondence of William James*, vol. 1 (Charlottesville: University Press of Virginia, 1992), 28–29.

7. See the explanation of "Counter-Irritation," in Richard D. Hoblyn, *A Dictionary of Terms Used in Medicine and the Collateral Sciences*, 9th ed. (London: Whittaker & Co., 1868), 178. It was this medical theory that formed the basis of James's supposition, on another occasion, that "a couple of very painful boils on [his] loins" were responsible for an improvement in his back ache! See William James to Henry James, June 1, 1869, Skrupskelis and Berkeley, eds., *Correspondence*, 1:78.

8. William James to Robertson James, December 26, 1869, Skrupskelis and Berkeley, eds., *Correspondence*, 4:394–95. In a letter to a doctor friend, in early 1869, James informed him: "I find a gentle stimulation with Iodine to be beneficial to me. But I suspect counterirritation, if done too *continuously*, does harm. It must be practised for a few days, and then omitted for a greater number." See William James to Henry Pickering Bowditch, January 24, 1869, ibid., 363.

9. William James to Robertson James, January 27, 1868, Skrupskelis and Berkeley, eds., *Correspondence*, 4:262. This letter provides a revealing counterpoint to the widely

accepted biographical narrative that James never had any genuine interest in practicing as a doctor.

10. William James to Henry James, July 10, 1868, Skrupskelis and Berkeley, eds., *Correspondence*, 1:53.

11. William James to Henry Pickering Bowditch, January 24, 1869; William James to Henry Pickering Bowditch, December 29, 1869, Skrupskelis and Berkeley, eds., *Correspondence*, 4:361, 397.

12. William James to Henry James, December 5, 1869; William James to Henry James, December 27, 1869, Skrupskelis and Berkeley, eds., *Correspondence*, 1:130, 132.

13. William James to Henry James, December 27, 1869, ibid., 132.

14. William James to Henry Pickering Bowditch, December 29, 1869, Skrupskelis and Berkeley, eds., *Correspondence*, 4:396.

15. See, for example, William James to Alice James, March 16, 1868; William James to Oliver Wendell Holmes, September 17, 1867, ibid., 200, 265.

16. William James to Oliver Wendell Holmes, September 17, 1867, ibid., 200.

17. William James to Henry James, May 7, 1870, Skrupskelis and Berkeley, eds., *Correspondence*, 1:158–59.

18. William James to Robertson James, April 28, 1870, Skrupskelis and Berkeley, eds., *Correspondence*, 4:406.

19. James, *The Varieties*, 136.

20. For James on evil, see Sami Pihlström, *Taking Evil Seriously* (New York: Palgrave Macmillan, 2014); Todd Lekan, "Pragmatist Moral Philosophy and Moral Life: Embracing the Tensions," in *The Jamesian Mind*, ed. Sarin Marchetti (London, New York: Routledge, 2022).

21. William James to Edgar Beach Van Winkle, March 1, 1858, Skrupskelis and Berkeley, eds., *Correspondence*, 4:12.

22. William James to Thomas Wren Ward, May 24, 1868, ibid., 308.

23. Henry James, "The Nature of Evil," in *Henry James Senior: A Selection of His Writings*, ed. Giles Gunn (Chicago: American Library Association, 1974), 150–51.

24. Diary 1 (1868–73), William James Papers (MS Am 1092.9 [4550]), Houghton Library, Harvard University, accessed July 13, 2022, https://hollisarchives.lib.harvard.edu/repositories/24/digital_objects/14160 (seq. 11–12).

25. William James to Henry James, April 5, 1868, Skrupskelis and Berkeley, eds., *Correspondence*, 1:41–43.

26. William James to Thomas Wren Ward, May 24, 1868, Skrupskelis and Berkeley, eds., *Correspondence*, 4:308.

27. Ibid., 309.

28. William James to Henry James, October 25 [1869], Skrupskelis and Berkeley, eds., *Correspondence*, 1:113–14. I have omitted the French citation and left in James's loose translation. The editors of the collected correspondence attribute the quotation to the French philosopher Blaise Pascal. Ibid., 114. See Blaise Pascal, *Les Pensées* (Paris: Alphonse Lemerre, 1777), 66.

29. William James, Miscellaneous notes, William James Papers (MS Am 1092.9 [4473]), Houghton Library, Harvard University.

30. William James to Henry James, May 7, 1870, Skrupskelis and Berkeley, eds., *Correspondence*, 1:159.

31. Marcus Dods, ed., *The Works of Aurelius Augustine, Bishop of Hippo*, vol. 9, *On Christian Doctrine; the Enchiridion; on Catechising; and on Faith and the Creed* (Edinburgh: T. & T. Clark, 1873), 182.

32. For Henry James Sr. the only significant evil was the "spiritual evil" that arises from man's obstinate "airs of self-sufficiency" and "separates the soul from God." James, "The Nature of Evil", 149-51.

33. William James to Robertson James, November 14, 1869, Skrupskelis and Berkeley, eds., *Correspondence*, 4:389-90.

34. James, Miscellaneous notes, William James Papers.

35. Frederick Rosen, *Classical Utilitarianism from Hume to Mill* (London: Routledge, 2003), 221. Collectively these three men have been deemed "the English Utilitarians." They were a "group of men who for three generations had a conspicuous influence upon English thought and political action." See Sir Leslie Stephen, *The English Utilitarians*, vol. 1, *Jeremy Bentham* (New York: Augustus M. Kelley, 1968), 1.

36. Rosen, *Classical Utilitarianism*, 220.

37. In a letter to Oliver Wendell Holmes, James wrote: "I know that the brightest jewel in the crown of Utilitarianism is that every notion hatched by the human mind receives justice & tolerance at its hands." See William James to Oliver Wendell Holmes, May 15, 1868, Skrupskelis and Berkeley, eds., *Correspondence*, 4:302-3. See also William James to Thomas Wren Ward, March 14, 1870, ibid., 404. There James writes, "Lecky's misunderstanding of the Utilitarian position, [has] been to my mind [a] very heavy blow against the value of all human testimony." This appears to be a reference to William Lecky's controversial introduction to his work, William Edward Hartpole Lecky, *A History of European Morals from Augustus to Charlemagne* (New York: D. Appleton & Company, 1869).

38. Andrew Pyle, *Utilitarianism: Key Nineteenth-Century Journal Sources* (London: Routledge, 1998), viii. See Halliwell and Rasmussen, eds., *William James and the Transatlantic Conversation*, and, in particular, Jeremy Carrette, "Growing Up Zig-Zag: Reassessing the Transatlantic Legacy of William James," in ibid.

39. James, *Pragmatism*, 32.

40. The connections between utilitarian ideas and James's philosophy have received some attention. See, for example, Christopher Hookway, "Logical Principles and Philosophical Attitudes: Peirce's Response to James's Pragmatism," in *The Cambridge Companion to William James*, ed. Ruth Anna Putnam (Cambridge: Cambridge University Press, 1997), 148-54; Graham H. Bird, "Moral Philosophy and the Development of Morality," in *The Cambridge Companion to William James*, ed. Putnam, 261; Piers H. G. Stephens, "James, British Empiricism, and the Legacy of Utilitarianism," in *The Jamesian Mind*, ed. Marchetti.

41. Robert Richardson, *William James in the Maelstrom of American Modernism* (Boston and New York: Houghton Mifflin Company, 2007), 155.

42. John Stuart Mill, *Auguste Comte and Positivism* (London: N. Trübner & Co., 1865), 134.

43. J. S. Mill, "Utilitarianism," in *Utilitarianism and Other Essays*, ed. Alan Ryan (London: Penguin Books, 1987), 305.

44. Mill, *Auguste Comte*, 146.

45. William James to Henry James Sr., October 28, 1867, Skrupskelis and Berkeley, eds., *Correspondence*, 4:221.

46. William James to Henry James Sr., December 26, 1867; William James to Thomas Wren Ward, November 7, 1867; William James to Thomas Wren Ward, January 7, 1868, ibid., 226, 243, 250.

47. Mill, *Auguste Comte*, 146-47.

48. William James to Henry James, February 12, 1868, Skrupskelis and Berkeley, eds., *Correspondence*, 1:31.

49. William James to Thomas Wren Ward, October 9, 1868; William James to Oliver Wendell Holmes, May 15, 1868, Skrupskelis and Berkeley, eds., *Correspondence*, 4:300–302, 345–47.

50. These undated notes were found in an envelope marked "Pomfret," a place that James and his family visited for a holiday on more than one occasion. Perry attributed these personal notes to the year 1869. Although James also spent some time there the following summer, of 1870, cross-referencing the content of the notes with his dated correspondence suggests that the "Pomfret notes" were written during the first visit.

51. Jeremy Bentham, *An Introduction to the Principles of Morals and Legislation*, ed. J. H. Burns and H. L. A. Hart (Oxford: Clarendon Press, 1996), 11.

52. Bentham was trained in the law, and, though he never practiced, his abiding passion was for legal reform. The dissatisfactions that he felt with English law stimulated his earliest writings, and his *Principles of Morals and Legislation* was intended as an introduction to a (never completed) model penal code. John Troyer, "Introduction," in *The Classical Utilitarians: Bentham and Mill*, ed. John Troyer (Indianapolis: Hackett Publishing Company, Inc., 2003), ix.

53. William James to Thomas Wren Ward, January 7, 1868, Skrupskelis and Berkeley, eds., *Correspondence*, 4:248.

54. For an examination of nineteenth-century understandings of the concepts of sympathy and altruism, see Rob Boddice, *The Science of Sympathy: Morality, Evolution and Victorian Civilization* (Urbana: University of Illinois Press, 2016); Susan Lanzoni, "Sympathy in Mind," *Journal of the History of Ideas* 70, no. 2 (2009): 265–87; Thomas Dixon, *The Invention of Altruism: Making Moral Meanings in Victorian Britain* (Oxford: Oxford University Press/British Academy, 2008).

55. William James to Henry James Sr., September 5, 1867, Skrupskelis and Berkeley, eds., *Correspondence*, 4:194.

56. James, Miscellaneous notes, William James Papers.

57. William James, Notebook 26, William James Papers (MS Am 1092.9 [4520]), Houghton Library, Harvard University.

58. Madame de Staël, *Réflexions sur le suicide* (London: L. Deconchy, 1813).

59. Jean Dumas, *Traité du suicide, ou, Du meurtre volontaire de soi-même* (Amsterdam: D. J. Changuion, 1773).

60. Johannes Robeck, *De morte voluntaira philosophorum et bonorum vivorum* (1735). Johannes Robeck is known as one of the few philosophers writing on this subject who acted on his conclusions and, after completing his manuscript, drowned himself at sea. See Georges Minois, *History of Suicide: Voluntary Death in Western Culture*, trans. Lydia G. Cochrane (Baltimore: John Hopkins Press, 1999), 220–21.

61. John Donne, *Biathanatos: A Declaration of that Paradoxe, or Thesis, that Selfe-homicide is not so naturally Sinne, that it may never be Otherwise* (1647). James's entry is brief ("Donne on S. 1644"), and appears to list the wrong publication date for Donne's *Biathanatos*, but there has always been confusion about the dates associated with this book owing to the long gap between completion and publication. (It was written around 1608 but published for the first time, and against the author's wishes, in 1647.) See Notebook 26, William James Papers; Michael Rudick and M. Pabst, "Introduction," to John Donne, *Biathanatos: A Modern-Spelling Critical Edition*, ed. Michael Rudick and M. Pabst Battin (New York: Garland Publishing Company, 1982), x, ix.

62. His treatise specifically about this subject was published along with another brief work under the title David Hume, *Essays on Suicide and the Immortality of the Soul* (Lon-

don, 1783). The text had also been published earlier, in 1770, in an anonymous French edition. See Minois, *History of Suicide*, 252.

63. Montaigne discusses suicide in several of his *Essays* (see Minois, *History of Suicide*, 88.), but James notes one in particular. His reference reads: "Montaigne. Essais liv. ii ch. XIII." which seems to be a reference to a section in Book II of the essays: "Chapter XIII: Of Judging of Other's Death." See James, Notebook 26, William James Papers. For the original French edition of the essays, see Michel de Montaigne, *Essais* (Bordeaux: Simon Millanges, Jean Richer, 1580).

64. Jean-Jacques Rousseau, *La nouvelle Héloïse* (Amsterdam: Marc-Michel Rey, 1761). The text is an epistolary novel, and James specifies a particular section, namely, letters 21–22. James, Notebook 26, William James Papers.

65. De Quincey was famous for his 1821 autobiographical text *Confessions of an English Opium-Eater* about the pleasures and pains of an opium habit. A later collection of his essays included one on the topic of suicide. See Thomas De Quincey, *The Note Book of an English Opium-Eater* (Boston: Ticknor and Fields, 1855).

66. For more detail about James's interest in Stoic philosophy, see Emma Sutton, "Marcus Aurelius, William James and the 'Science of Religions,'" *William James Studies* 4 (2009).

67. See Chauncey Wright to Grace Norton, July 18, 1875, cited in Perry, *Thought and Character*, 1:531.

68. Minois, *History of Suicide*, 43. There were exceptions to this monolithic standpoint including Donne's *Biathanatos*, which is written from a Christian perspective but espoused, on extremely limited terms, the legitimacy of suicide. See Donne, *Biathanatos*.

69. Minois, *History of Suicide*, 46–48.

70. Michael MacDonald and Terrence Murphy, *Sleepless Souls: Suicide in Early Modern England*, Oxford Studies in Social History, paperback ed. (Oxford: Clarendon Press, 1993), 17.

71. Minois, *History of Suicide*, 43.

72. MacDonald and Murphy, *Sleepless Souls*, 17.

73. Seneca, *Seneca's Letters to Lucilius*, vol. 1, trans. E. Phillips Barker (Oxford: Clarendon Press, 1932), 82. Cited in Minois, *History of Suicide*, 52.

74. Minois, *History of Suicide*, 54.

75. Marcus Aurelius, *The Meditations of Marcus Aurelius Antoninus*, ed. R. B. Rutherford, trans. A. S. L. Farquharson (Oxford: Oxford University Press, 1989), 42. Cited in Minois, *History of Suicide*, 54.

76. William James to Thomas Wren Ward, January 7, 1868, Skrupskelis and Berkeley, eds., *Correspondence*, 4:250.

77. Many of the texts referenced by James were written by French authors and centered on the implications and explanations of what was perceived as a dramatic increase in the number of suicides in France during the middle decades of the nineteenth century. See, for example, Louis Bertrand, *Traité du suicide: Considéré dans ses rapports avec la Philosophie, la Théologie, la Médecine et la Jurisprudence* (Paris: J. B. Baillière, 1857); J. B. Petit, "Recherches statistiques de l'étiologie du suicide" (Paris, 1850); Gustave-François Étoc-Demanzy, *Recherches statistiques sur le suicide, appliques à l'hygiene publique et à la medicine legale* (Paris: Germer-Bailliere, 1844).

78. Forbes Winslow, *The Anatomy of Suicide* (London: Henry Renshaw, 1840), v.

79. Sir Matthew Hale, *The History of the Pleas of the Crown*, vol. 2, ed. Sollom Emlyn (London, 1736), 29–32.

80. Joel Eigen, *Witnessing Insanity: Madness and Mad-Doctors in the English Court* (New Haven, CT: Yale University Press, 1995), 158.

81. In his notebook James lists simply "Bertrand," but this is most likely a reference to Bertrand, *Traité du suicide*.

82. "Bertrand on Suicide," *American Journal of Insanity* 14 (1857). For a brief reference to the position maintained by Bertrand, see W. S. F. Pickering and Geoffrey Walford, *Durkheim's Suicide: A Century of Research and Debate* (London: Routledge, 2000), 72.

83. In his *On Obscure Diseases*, for example, Winslow cites the case of a boy whose "whole moral character was found to have undergone a complete metamorphosis" after he sustained a severe injury. "From being a well-conditioned boy, kind and affectionate to his parents, steady in his habits, sober, of unimpeachable veracity, he became a drunkard, liar, and thief." He goes on to emphasize that while he was "clever, intelligent, sharp-witted, . . . his every action was perfectly brutal." Forbes Winslow, *On Obscure Diseases of the Brain, Disorders of the Mind: Their Incipient Symptoms, Pathology, Diagnosis, Treatment, and Prophylaxis* (Philadelphia: Blanchard and Lea, 1860), 159.

84. Jonathan Andrews, "Winslow, Forbes Benignus (1810–1874)," in *Oxford Dictionary of National Biography* (Oxford: Oxford University Press, 2004).

85. James, *The Varieties*, 134. The description of the vision was originally attributed to an anonymous Frenchman when it was published in *The Varieties of Religious Experience*, but James later confessed to his translator that "the document . . . is my own case—acute neurasthenic attack with phobia. I naturally disguised the *provenance!*" William James to Frank Abauzit, June 1, 1904, cited in *The Varieties*, ed. Burkhardt, Bowers, and Skrupskelis, 508.

86. William James to Katharine James Prince, December 13, 1863, Skrupskelis and Berkeley, eds., *Correspondence*, 4:87.

87. The characteristic pencil marginalia take the form of lines drawn in the margin of the pages to highlight various passages and a brief personal index of significant pages written inside the back cover. Eugene Taylor describes the prevalence of this kind of marking up of texts by James in his comprehensive reconstruction of James's 1896 Lowell Lectures. See Taylor, *Exceptional Mental States*, xi. In this instance the pencil marginalia refer to a passage in the chapter entitled "Confessions of Patients after Recovering from Insanity; or The Condition of the Mind When in a State of Aberration." These markings appear, however, to date from a later rereading of the volume, some years after the letter to his cousin. In his younger years James practiced different systems of annotation. In some of the volumes he purchased and read during his European wanderings in the late 1860s, he made notes or annotations on a separate piece of paper and stuck the edges of these directly onto the relevant pages so that they formed additional, handwritten leaves interwoven with the original text. See, for example, James's copies of Rudolph Hermann Lotze, *Medicinische Psychologie oder Physiologie der Seele* (Leipzig: Wiedmann'sche Buchhandlung, 1852); J. Luys, *Recherches sur le Système Nerveux Cérébro-Spinal sa Structure, ses Fonctions, ses Maladies* (Paris: J.-B. Bailére et Fils, 1865). Such notes are not to be found in his copy of Winslow's book, which he read, originally, several years earlier still in 1863.

88. Winslow, *On Obscure Diseases of the Brain*, 150–51, 248–49, 492–93, 528–29.

89. Ibid., 13.

90. For a retrospective description of his experiences of "ennui" as a "lonely invalid," during his travels around Europe in the 1860s, see William James to Henry James, May 25, 1873, Skrupskelis and Berkeley, eds., *Correspondence*, 1:209.

91. Winslow, *On Obscure Diseases*, 248.

92. Ibid., 248–49.

93. Ian Dowbiggin and Jan Goldstein make the case that there were professional agendas at stake in treatises about the etiology and treatment of insanity. See Ian R. Dowbiggin, *Inheriting Madness: Professionalisation and Psychiatric Knowledge in Nineteenth Century France* (Berkeley: University of California Press, 1991); Jan Goldstein, *Console and Classify: The French Psychiatric Profession in the Nineteenth Century* (Chicago: University of Chicago Press, 1987). A contemporary commentator and spiritualist philosopher, Albert Lemoine, writing in 1862, stated that a doctor was essentially a "healer of the corporeal machine." As such his expertise was essentially redundant should insanity be a purely mental derangement, unconnected with bodily illness. Dowbiggin, *Inheriting Madness*, 49. Some nineteenth-century doctors did, however, believe it was possible to maintain a philosophical compromise that rationalized a reliance on moral (psychological) treatment, or explanatory causes, while still supporting the physicalist claims of the physiologists. See, for example, François Leuret, *Du traitement moral de la folie* (Paris: J.-B. Baillière 1840).

94. Henry Maudsley, *Body and Mind: An Inquiry into their Connection and Mutual Influence, Specially in Reference to Mental Disorders* (New York: D. Appleton and Company, 1871), 68.

95. William James to Thomas Wren Ward, March 1869, Skrupskelis and Berkeley, eds., *Correspondence*, 4:370.

96. Henry James, ed., *The Letters of William James*, vol. 1 (London: Longmans, Green, and Co., 1920), 152. Wright would not have considered himself "a materialist," however. He was vociferous in his insistence that mental events are "fully *conditioned* by physical ones but are not *analyzable* into physical atoms as parts," as the doctrine of materialism maintains. See Philip P. Wiener, "Chauncey Wright's Defence of Darwin and the Neutrality of Science," *Journal of the History of Ideas* 6, no. 1 (January 1945): 35. As Jan Goldstein has pointed out, however, such metaphysical hairsplitting was commonplace. Essentially, nobody described *themselves* as "materialist" since the term had a "highly pejorative sound" during the nineteenth century. See Goldstein, *Console and Classify*, 244.

97. Croce, *Science and Religion*, 151, 162–63.

98. John Fiske remarks that Wright was "somewhat deficient in the literary knack of expressing his thoughts in language generally intelligible or interesting [and] singularly devoid of the literary ambition which leads one to seek to influence the public by written exposition." John Fiske, "Chauncey Wright," in *The Evolutionary Philosophy of Chauncey Wright*, vol. 3, *Influence and Legacy*, ed. Frank X. Ryan (Bristol: Thoemmes Press, 2000), 6.

99. Chauncey Wright, "McCosh on Tyndall," in *The Evolutionary Philosophy*, ed. Ryan, 381.

100. William James, "Chauncey Wright," in ibid., 2. This subtext is also present in a passing comment, made by James, in a letter to his family in 1872. He mentions having recently spent time with "the great Chauncey Wright who now lies (as to his system of the universe) in my arms as harmless as a babe." William James to Alice James, Henry James, Catharine Walsh, May 30 [1872], Skrupskelis and Berkeley, eds., *Correspondence*, 4:422.

101. See William James, "Miscellanea 1: Mostly Concerning Empiricism," in *The Works of William James: Manuscript Essays and Notes*, ed. Frederick H. Burkhardt, Fredson Bowers, and Ignas K. Skrupskelis (Cambridge, MA: Harvard University Press, 1988), 133–42; William James, "Draft on Brain Processes and Feelings," in ibid., 247–56.

102. James's interest in alienism originated early on in his student days. Diary entries,

discussing, for example, the work of the French doctor Jacques-Joseph Moreau de Tours, testify to his ongoing engagement with the literature on this subject. Diary 1, William James Papers, online access (seq. 74).

103. William James, "Contributions to Mental Pathology, by Isaac Ray (1873)," in *The Works of William James: Essays, Comments, and Reviews*, ed. Frederick H. Burkhardt, Fredson Bowers, and Ignas K. Skrupskelis (Cambridge, MA: Harvard University Press, 1987), 267–69; William James, "Recent Works on Mental Hygiene (1874)," in ibid., 276–79; William James, "Two Reviews of *Principles of Mental Physiology*, by William B. Carpenter (1874)," in ibid., 275.

104. Henry James Sr. to Henry James, March 18, 1873, William James Papers (MS Am 1092.9 [4199]), Houghton Library, Harvard University.

105. James, "Two Reviews," 271–73.

106. William James, "*Grundzüge der physiologischen Psychologie* by Wilhelm Wundt (1875)," in *Essays, Comments, and Reviews*, ed. Burkhardt, Bowers, and Skrupskelis, 302. Robert Richards discusses James's development of this argument in Richards, "James's Uses of Darwinian Theory."

107. William James, "Lowell Lectures on 'The Brain and the Mind' (1878)," in *Manuscript Lectures*, ed. Burkhardt, Bowers, and Skrupskelis, 24, 25, 28.

108. William James to Shadworth Hollway Hodgson, March 11 [1879], Skrupskelis and Berkeley, eds., *Correspondence*, 5:44. For further exploration of James's stance on epiphenomenalism, see Alexander Klein, "William James's Objection to Epiphenomenalism," *Philosophy of Science* 86, no. 5 (2019); Alexander Klein, "James on Consciousness," in *Oxford Handbook*, ed. Klein, accessed July 7, 2022, https://www.oxfordhandbooks.com /view/10.1093/oxfordhb/9780199395699.001.0001/oxfordhb-9780199395699-e-4.

109. It is important to note that, for James, proposing an active role for consciousness, in opposition to the assumptions of medical materialism, did not imply a belief in free will: "Free-will is in short, no necessary corollary of giving causality to consciousness." William James to James Jackson Putnam, January 17, 1879, Skrupskelis and Berkeley, eds., *Correspondence*, 5:34.

110. According to Perry, James experienced a "spiritual crisis" that he characterizes as "the ebbing of the will to live, for lack of a philosophy to live by—a paralysis of action occasioned by a sense of moral impotence." Perry, *Thought and Character*, 1:322.

111. A well-known Jamesian diary entry, for April 1870, records the impression that reading Renouvier's philosophy made on him: "I finished reading the first part of Renouvier's 2nd Essays and see no reason why his definition of free will—'the sustaining of a thought *because I choose to* when I might have other thoughts'—need be the definition of an illusion. . . . My first act of free will shall be to believe in free will." Diary 1, William James Papers, online access (seq. 85).

112. William James to Francis Herbert Bradley, January 3, 1898, Skrupskelis and Berkeley, eds., *Correspondence*, 8:333.

113. William James, "The Dilemma of Determinism," in *The Will to Believe*, ed. Burkhardt, Bowers, and Skrupskelis, 136–37.

114. Although James appears to have been consistent in his support for the physiological explanations of criminality, his opinions on the implications of this perspective for punishment and public policy were less clear-cut and changed over time. William James, "*Responsibility in Mental Disease*, by Henry Maudsley (1874)," in *Essays, Comments, and Reviews*, ed. Burkhardt, Bowers, and Skrupskelis; "William James, 'The Jukes,' by Richard L. Dugdale (1878)," in ibid.

115. Leslie Stephen to William James, October 16, 1884, Skrupskelis and Berkeley, eds., *Correspondence*, 5:528–29.

116. In his lectures on *Pragmatism* that he published many years later, James defined the philosophy of free will as a "cosmological theory of *promise*" that is only needed by "persons in whom knowledge of the world's past has bred pessimism." Its appeal is that "it holds up improvement as at least possible." William James, *Pragmatism*, 61. For an in-depth discussion of this theme, see Mathew A. Foust, "William James and the Promise of *Pragmatism*," *William James Studies* 2 (2007), accessed July 1, 2021, https://william jamesstudies.org/william-james-and-the-promise-of-pragmatism/.

117. Diary 1, William James Papers, online access (seq. 87).

118. William James to Thomas Wren Ward, October 9 [1868], Skrupskelis and Berkeley, eds., *Correspondence*, 4:347.

119. William James to Henry James, April 23, 1869; William James to Henry James, June 1, 1869, Skrupskelis and Berkeley, eds., *Correspondence*, 1:67, 78.

120. William James to Robertson James, July 25, 1870, Skrupskelis and Berkeley, eds., *Correspondence*, 4:409.

121. Ibid. The specific section is *On Providence, II*. It seems likely that James read a French translation of the text in question, as his own words follow those of this Latin-French edition closely: M. Nisard, ed., *Oeuvres Complètes de Sénèque le Philosophe, avec la Traduction en Français* (Paris: J. J. Dubochet et Compagnie, 1844), 127.

122. William James, "Lowell Lectures (1878)," 27.

123. This contemporary context is made explicit in James, *The Will to Believe*.

124. William James, "Rationality, Activity and Faith," 59, 64.

125. Ibid., 73–74.

126. Ibid., 79.

127. Ibid.

128. Ibid.

129. Ibid., 80n.

130. Ibid., 79–80.

131. James, "The Dilemma of Determinism," 138.

132. James, "Rationality, Activity and Faith," 80.

133. Ibid., 84.

134. William James, "Are We Automata?," in *The Works of William James: Essays in Psychology*, ed. Frederick H. Burkhardt, Fredson Bowers, and Ignas K. Skrupskelis (Cambridge, MA: Harvard University Press, 1983), 38.

135. William James to Charles Renouvier, September 28, 1882, Skrupskelis and Berkeley, eds., *Correspondence*, 5:260.

Chapter Two

1. James described Beard to his wife as "an able & honest, tho' I fear somewhat conceited man." William James to Alice Howe Gibbens James [June 24, 1879], Skrupskelis and Berkeley, eds., *Correspondence*, 5:54.

2. The long list of symptoms that the American neurologist listed as typical of the disease of neurasthenia included, but were not limited to, tenderness of the scalp, dilated pupils, sick headaches and various forms of head pain, noises in the ear, deficient mental control, mental irritability, hopelessness, morbid fear, tenderness of teeth and gums, nervous dyspepsia, desire of stimulants and narcotics, abnormalities of the secretions,

sweating hands and feet, salivation, tenderness of the spine, heaviness of the loins and limbs, sensitiveness to changes in the weather, and ticklishness. See George M. Beard, *A Practical Treatise on Nervous Exhaustion (Neurasthenia): Its Symptoms, Nature, Sequences, Treatment* (New York: William Wood & Company, 1880), 11–85.

3. Massachusetts Department of Public Health, *State Board of Health of Massachusetts: A Brief History of Its Organization and Its Work, 1869–1912* (Boston: Wright & Potter, 1912), 8–9.

4. James C. Whorton, *Crusaders for Fitness: The History of American Health Reformers* (Princeton, NJ, and Guildford: Princeton University Press, 1982), 14–15.

5. Charles E. Rosenberg, "Catechisms of Health: The Body in the Prebellum Classroom," *Bulletin of the History of Medicine* 69, no. 2 (1995): 183.

6. Elisha Bartlett, *Obedience to the Laws of Health, a Moral Duty: A Lecture Delivered before the American Physiological Society, January 30, 1838* (Boston: Julius A. Noble, 1838).

7. Herbert Spencer, *Education: Intellectual, Moral and Physical* (New York: D. Appleton and Company, 1861), 282–83.

8. Whorton, *Crusaders for Fitness*, 29–33.

9. See, for example, a calculation of the "million and a quarter dollars" cost to society associated with the extended "Jukes" family, which James highlighted in his review of Dugdale's book on "the descendants of the notorious 'Margaret, mother of criminals.'" (The Jukes was a pseudonym.) James, "The Jukes," 345–46. See also James, "Fifth Annual Report."

10. Spencer cites the role of the "physical sins" of our "forefathers'" in bringing about ill health. Spencer, *Education*, 41.

11. John Waller traces the emergence of discussions that extrapolated concerns about individual lineages and applied them to the nation and race. John C. Waller, "Ideas of Heredity, Reproduction and Eugenics in Britain, 1800–1875," *Studies in the History and Philosophy of Biology and Bio-medical Sciences* 32, no. 3 (2001): 457–89.

12. Ibid., 458.

13. James, "Recent Works," 277.

14. Ibid., 277–78.

15. Ibid., 278.

16. See, for example, Scottish doctor Thomas Clouston's thoughts on women's education in T. S. Clouston, *Female Education from a Medical Point of View* (Edinburgh: Macniven & Wallace, 1882). Analyses of James and gender issues include Kim Townsend, *Manhood at Harvard: William James and Others* (New York: W. W. Norton & Company, 1996); Erin C. Tarver and Shannon Sullivan, eds., *Feminist Interpretations of William James* (University Park: Pennsylvania State University Press, 2015.).

17. James, "Fifth Annual Report," 281.

18. Ibid., 281–82.

19. William James to Alice Howe Gibbens James [September 1876], Skrupskelis and Berkeley, eds., *Correspondence*, 4:545.

20. Ibid., 545–46.

21. William James, "Physiology for Practical Use, ed. by James Hinton (1874)," in *Essays, Comments, and Reviews*, ed. Burkhardt, Bowers, and Skrupskelis, 284–85.

22. Skrupskelis, "Introduction," in *Manuscript Lectures*, ed. Burkhardt, Bowers, and Skrupskelis, xxiv.

23. Joseph W. Warren, "Alcohol Again: A Consideration of Recent Misstatements of Its Physiological Action," reprinted from *Boston Medical and Surgical Journal* (Boston: Cupples and Hurd, July 7, 14, 1887); G. Bunge, "Die Alkoholfrage, Ein Vortrag" (Basel:

L. Reinhardt, 1894); August Forel, "Die Trinklitten, ihre hygienische und sociale Bedeutung. Ihre Beziehungen zur Akademischen Jugend" (Basel: L. Reinhardt, 1894); Daniel Merriman, "A Sober View of Abstinence," in *Bibliotheca Sacra* (Andover: Warren F.
Draper, October 1881); "The Physiological Influence of Alcohol," in *Edinburgh Review*
(n.d.); Geo C. Kingsbury, "Alcohol: Its Use and Misuse," in *The Humanitarian* (n.d.); Joseph W. Warren, "The Effect of Pure Alcohol on the Reaction Time, with a Description
of a New Chronoscope," *Journal of Physiology* 8, no. 6 (1887); Robert Koppe, "Das Alkoholsiechthem und Die Kurzlebigkeit des modernen Menschengeschlechts" (Moskau:
E. Liessner & J. Romahn, 1894).

24. Holly Berkley Fletcher examines the sociological and Philip McGowan the physiological arguments mobilized in support of this stance. Holly Berkley Fletcher, *Gender
and the American Temperance Movement of the Nineteenth Century* (London: Routledge,
2007); Phillip McGowan, "AA and the Redeployment of Temperance Literature," *Journal of American Studies* 48, no. 1 (2014).

25. William James, "Scientific View of Temperance (1881)," in *Essays, Comments, and
Reviews*, ed. Burkhardt, Bowers, and Skrupskelis, 19.

26. One set of James's lecture notes makes reference, in connection with this topic,
to "Lallemand & Co." For more detail about this specific debate, and the findings of
MM. Lallemand, Perrin, and Duroy, see a contemporary account in J. C. Langmore and
William Roberts, "Reports of Societies," *British Medical Journal*, no. 1 (1865): 494–96.

27. James, "Scientific View of Temperance," 20.

28. Ibid., 20–21.

29. Ibid., 21.

30. Ibid.

31. William James, "Notes for a Lecture on 'The Physiological Effects of Alcohol'
(1886)," in *Manuscript Lectures*, ed. Burkhardt, Bowers, and Skrupskelis, 43, 45; William
James, "Draft and Outline of a Lecture on 'The Effects of Alcohol' (1895)," in ibid., 50.

32. James, "Scientific View of Temperance," 20.

33. James, "Draft and Outline," 50.

34. William James to Mary Robertson Walsh James, June 12, 1867, Skrupskelis and
Berkeley, eds., *Correspondence*, 4:176.

35. James, "Draft and Outline," 50.

36. James, "Scientific View of Temperance," 20–21.

37. See, for example, Berkley Fletcher, *Gender and the American Temperance Movement*; Whorton, *Crusaders for Fitness*; Joseph Gusfield, *Symbolic Crusade: Status Politics and the American Temperance Movement* (Urbana and London: University of Illinois
Press, 1963); John J. Rumbarger, *Profits, Power, and Prohibition: Alcohol Reform and the
Industrializing of America, 1800–1930* (Albany: State University of New York Press, 1989).

38. For a nuanced account of this moment, see chapter 4 of Fletcher, *Gender and the
American Temperance Movement*.

39. William James to Alice Howe Gibbens James, February 3 [1888]; William James
to Alice Howe, February 5, 1888, Ignas K. Skrupskelis and Elizabeth M. Berkeley, eds.,
The Correspondence of William James, vol. 6 (Charlottesville: University Press of Virginia, 1998), 307–8.

40. William James to Robertson James, May 26, 1879, Skrupskelis and Berkeley, eds.,
Correspondence, 5:53.

41. William James to Mary Holton James, October 4, 1884, ibid., 527.

42. William F. Bynum, "Chronic Alcoholism in the First Half of the 19th Century,"
Bulletin of the History of Medicine 42, no. 2 (1968): 161.

43. James, "The Jukes," 346.

44. Bynum, "Chronic Alcoholism," 171.

45. James, "Draft and Outline," 51.

46. Timothy A. Hickman, "'Mania Americana': Narcotic Addiction and Modernity in the United States, 1870–1920," *Journal of American History* 90, no. 4 (2004): 1281.

47. James, "Draft and Outline," 50–51.

48. James, "Notes for a Lecture," 45.

49. See, for example, Joseph M. Thomas, "Figures of Habit in William James," *New England Quarterly* 66, no. 1 (1993); Renee Tursi, "William James's Narrative of Habit," *Style* 33, no. 1 (1999); Michael S. Lawlor, "William James's Psychological Pragmatism: Habit, Belief and Purposive Human Behaviour," *Cambridge Journal of Economics*, no. 30 (2006). George Cotkin and Gerald Myers also contribute short analyses of James's interest in the concept of habit. See Cotkin, *William James*, 69–70; Gerald E. Myers, *William James: His Life and Thought* (New Haven, CT: Yale University Press, 1986), 199.

50. Perry, *Thought and Character*, 2:90.

51. Lesley Hall, "'It Was Affecting the Medical Profession': The History of Masturbatory Insanity Revisited," *Paedagogica Historica* 39, no. 6 (2003).

52. William James to Alexander Robertson James [October 6, 1900], Ignas K. Skrupskelis and Elizabeth M. Berkeley, eds., *The Correspondence of William James*, vol. 9 (Charlottesville: University of Virginia Press, 2001), 333–34.

53. All of these activities had potential health consequences according to Hinton's book on physiology and hygiene, which James reviewed favorably. See James Hinton, ed., *Physiology for Practical Use: By Various Writers*, 2 vols. (London: Henry S. King & Co., 1874).

54. Rosenberg, "Catechisms of Health," 192. The emphasis in the hygiene texts on the importance of "evacuation" for general health explains, in part, the seriousness with which the James brothers approached their problems with constipation. James Whorton, in his study of hygiene and constipation in the nineteenth century, describes how doctors warned that costiveness could lead to "a diseased stomach, . . . enlarged and diseased liver, . . . loss of memory, headache, heart disease, bleeding of the lungs, . . . falling of the womb, dyspepsia, piles, hectic fever, consumption." James C. Whorton, *Inner Hygiene: Constipation and the Pursuit of Health in Modern Society* (Oxford: Oxford University Press, 2000), 1.

55. William James to Alice Howe Gibbens James, September 24, 1882, Skrupskelis and Berkeley, eds., *Correspondence*, 5:255.

56. William James, *The Principles of Psychology*, vol. 1, ed. Frederick H. Burkhardt, Fredson Bowers, and Ignas K. Skrupskelis, The Works of William James (Cambridge, MA: Harvard University Press, 1981), 110–11.

57. Ibid., 111.

58. Robert Thomson, "Introduction," in William B. Carpenter, *Principles of Mental Physiology, with Their Applications to the Training and Discipline of the Mind, and the Study of Its Morbid Conditions*, 4th (1876) ed. (London: Routledge/Thoemmes Press, 1993), v.

59. James, "Two Reviews," 275; James, "Recent Works," 278.

60. William B. Carpenter, *Principles of Mental Physiology, with Their Applications to the Training and Discipline of the Mind, and the Study of Its Morbid Conditions* (New York: D. Appleton & Company, 1874), 341, 343.

61. Ibid., 344.

62. In his survey of the history of mental hygiene, Albert Deutsch states that, ordi-

narily, the term is associated with "*organized* efforts to promote mental health, more specifically, with the mental hygiene movement." Albert Deutsch, "The History of Mental
Hygiene," in *One Hundred Years of American Psychiatry*, ed. J. K. Hall, Gregory Zilboorg,
and Henry Alden Bunker (New York: Columbia University Press, 1944), 332. James's role
in the mental hygiene movement, which began in the early decades of the twentieth century, is discussed in chapter 5, "Politics and Pathology."

63. James, "Contributions," 267.

64. I. Ray, *Mental Hygiene* (Boston: Ticknor and Fields, 1863), 15.

65. According to J. M. Quen this term appeared for the first time in W. Sweetser, *Mental Hygiene, or An Examination of the Intellect and Passions, Designed to Illustrate Their
Influence on Health and the Duration of Life* (New York: J. & H. G. Langley, 1843). See
J. M. Quen, "Isaac Ray and Mental Hygiene in America," *Annals of the New York Academy of Sciences*, no. 291 (1977): 84, 86.

66. In later English language editions, the title was translated as *Hygiene of the Mind*.
See, for example, Baron Ernst von Feuchtersleben, *Hygiene of the Mind (Zur Diatetik der
Seele)*, trans. F. C. Sumner, 3rd (1910) ed. (New York: Macmillan Co., 1933).

67. Ernest von Feuchtersleben, *The Dietetics of the Soul*, 7th ed. (New York: C. S.
Francis & Co., 1854), 9, 17.

68. William James, "*Responsibility in Mental Disease*, by Henry Maudsley (1874)," in
Essays, Comments, and Reviews, ed. Burkhardt, Bowers, and Skrupskelis, 283.

69. Henry Maudsley, *Responsibility in Mental Disease* (London: Henry S. King & Co.,
1874), 300. For a survey of the concept of the will in a variety of nineteenth-century intellectual contexts, see Roger Smith, *Free Will and the Human Sciences in Britain, 1870–
1910* (London: Pickering & Chatto, 2013).

70. James, "*Responsibility in Mental Disease*," 283–84.

71. See chapter 1, "Misery and Metaphysics."

72. James, "*Responsibility in Mental Disease*," 283.

73. James, "Recent Works," 278.

74. Carpenter, *Principles of Mental Physiology*, 657–58.

75. Ibid., 658–66.

76. Ibid., 671, 673–75.

77. Carpenter, *Principles of Mental Physiology*, 674.

78. James, "Recent Works," 276, 278.

79. Carpenter, *Principles of Mental Physiology*, 345.

80. James, "Two Reviews," 275; James, "Recent Works," 279.

81. Gerald E. Myers, "Introduction," in William James, *Talks to Teachers on Psychology and to Students on Some of Life's Ideals*, ed. Frederick H. Burkhardt, Fredson Bowers, and Ignas K. Skrupskelis, The Works of William James (Cambridge, MA: Harvard
University Press, 1983), xi–xii.

82. William James to George Holmes Howison, October 27, 1897, Skrupskelis and
Berkeley, eds., *Correspondence*, 8:319.

83. James, *Talks to Teachers*, 14–16.

84. William James to William Torrey Harris, November 14, 1891, *The Correspondence
of William James*, vol. 7 (Charlottesville: University of Virginia Press, 1999), 220.

85. William James to Rosina Hubley Emmet, July 16, 1895, Skrupskelis and Berkeley,
eds., *Correspondence*, 8:55.

86. George Fielding Blandford, "Prevention of Insanity (Prophylaxis)," in *A Dictionary of Psychological Medicine: Giving the Definition, Etymology, and Synonyms of the*

Terms Used in Medical Psychology with the Symptoms, Treatment, and Pathology of Insanity and the Law of Lunacy in Great Britain and Ireland, ed. D. Hack Tuke (London: J. & A. Churchill, 1892), 998.

87. S. Weir Mitchell, *Wear and Tear, or Hints for the Overworked*, 8th ed. (Philadelphia: J. B. Lippincott Company, 1897).

88. Ray, *Mental Hygiene*, 121–22.

89. For a discussion of James's biological rhetoric within his *Talks to Teachers*, see Trevor Pearce, *Pragmatism's Evolution: Organism and Environment in American Philosophy* (London and Chicago: University of Chicago Press, 2020), 304–5.

90. James, *Talks to Teachers*, 24–25.

91. Ibid., 27.

92. Ibid., 26.

93. William James, "Appendix 1: Notes and Draft for Talks to Teachers," in *Talks to Teachers*, ed. Burkhardt, Bowers, and Skrupskelis, 185.

94. Ibid., 185.

95. Spencer, *Education*, 33, 39.

96. Herbert Spencer, *The Principles of Psychology* (London: Longman, Brown, Green, and Longmans, 1855), 436.

97. Charles Eliot, "Introduction," in Herbert Spencer, *Essays on Education and Kindred Subjects* (New York: E. P. Dutton & Co., 1911), viii.

98. Spencer, *Education*, 63.

99. James, *Talks to Teachers*, 48.

100. Ibid., 53.

101. Ibid., 111.

102. James, "Responsibility in Mental Disease," 283.

103. Maudsley, *Responsibility*, 300.

104. James, *Talks to Teachers*, 52–53.

105. Ibid., 67–68.

106. Like Carpenter and Maudsley, Clouston believed this faculty of self-control to be an important organizing principle within his model of insanity: "It is the characteristic of most forms of mental disease for self-control to be lost." In some types of insanity involving uncontrollable impulses to commit violence, drink, etc., "the loss of the power of inhibition is the chief and by far the most marked symptom." T. S. Clouston, *Clinical Lectures on Mental Diseases* (London: J. & A. Churchill, 1883), 310–18. Cited in William James, *The Principles of Psychology*, vol. 2, ed. Frederick H. Burkhardt, Fredson Bowers, and Ignas K. Skrupskelis, The Works of William James (Cambridge, MA: Harvard University Press, 1981), 1146–47.

107. James, *Talks to Teachers*, 112.

108. In their earliest incarnation James's lectures on pedagogical psychology, delivered in 1891 in the new Department of Pedagogy at Harvard, included a lecture devoted specifically to the subject of "Inhibition." This lecture is missing from the published *Talks to Teachers* but was summarized in the *Journal of Education* where the original lecture series was reported in some detail. Their reviewer noted that James stated that there are two types of inhibition: by "repression" and by "substitution." He characterized the former as "the older way; pent up energy, inward strain, fatigue afterwards." John Pierce, "Appendix 2: John Pierce's Summaries in the Journal of Education of James's Lectures," in *Talks to Teachers*, ed. Burkhardt, Bowers, and Skrupskelis, 217.

109. James, *Talks to Teachers*, 107–8.

110. In at least one edition of *Talks to Teachers* this is the only one of the "Talks to

Students" to be included in the volume. See William James, *Talks to Teachers on Psychology* (New York: Henry Holt and Company, 1946).

111. William James, "The Gospel of Relaxation," in *Talks to Teachers*, ed. Burkhardt, Bowers, and Skrupskelis, 117.

112. Ibid.

113. William James, "What Is an Emotion?," in *The Works of William James: Essays in Psychology*, ed. Frederick H. Burkhardt, Fredson Bowers, and Ignas K. Skrupskelis (Cambridge, MA: Harvard University Press, 1983), 170.

114. William James to Frederick George Bromberg, June 30, 1884, Skrupskelis and Berkeley, eds., *Correspondence*, 5:505. James's theory of the emotions has been the topic of much consideration. See, for example, Gerald E. Myers, "William James's Theory of Emotion," *Transactions of the Charles S. Peirce Society* 5, no. 2 (1969); Phoebe C. Ellsworth, "William James and Emotion: Is a Century Worth of Fame Worth a Century of Misunderstanding?," *Psychological Review* 101, no. 2 (1994); Thomas Dixon, *From Passions to Emotions: The Creation of Secular Psychological Category* (Cambridge: Cambridge University Press, 2003); Matthew Ratcliffe, "William James on Emotion and Intentionality," *International Journal of Philosophical Studies* 13, no. 2 (2005).

115. James, "The Gospel," 118.

116. Ibid., 122–24. Canadian educator James McLellan made a related point in his text: "The health of the body as a whole seems to be intimately connected with the *emotional* condition. The organic or common sensations coming from every part of the body, form, it is probable, the underlying emotional back-ground or disposition, and every disturbance of the health of the organism is reflected in a disturbance of the emotional attitude." J. A. McLellan, *Applied Psychology: An Introduction to the Principles and Practice of Education* (Toronto: Copp, Clark Company, Ltd., 1890), 157.

117. James, "The Gospel," 123, 126.

118. William James, "*Power through Repose*, by Annie Payson Call (1891)," in *Essays, Comments, and Reviews*, ed. Burkhardt, Bowers, and Skrupskelis, 419. See Annie Payson Call, *Power through Repose* (London: Gay and Hancock, 1914).

119. James, "The Gospel," 119-120.

120. William James to Margaret Mary James, June 30, 1900, Skrupskelis and Berkeley, eds., *Correspondence*, 9:243.

121. James, "Recent Works," 276.

122. Ibid., 277.

123. Ibid.

124. James, "The Gospel," 120–21.

125. Bruce Haley, *The Healthy Body and Victorian Culture* (Cambridge, MA: Harvard University Press, 1978), 21.

126. Ray, *Mental Hygiene*, 17, 54.

127. James does not refer to Benedikt by name, describing him only as a "Viennese neurologist of considerable reputation." Grace Foster and Donald Capps identify this neurologist as Sigmund Freud, but the term "Binnenleben" is specific to Benedikt's writings. See Grace Foster, "The Psychotherapy of William James," *Psychoanalytic Review* 32 (1945); Donald Capps, "Relaxed Bodies, Emancipated Minds, and Dominant Calm," *Journal of Religion and Health* 48, no. 3 (2009). For a discussion of Benedikt's ideas, see Oliver Somburg and Holger Steinberg, "Der Begriff des Seelen-Binnenlebens von Moriz Benedikt," in *Schriftenreihe der Deutschen Gesellschaft für Geschichte der Nervenheilkunde*, ed. W. J. Bock and B. Holdorff (2006). James had been following Benedikt's work from as early as 1869 as a letter to his brother Henry testifies. There James recommends Benedikt,

who was an expert on electro-therapeutics, as "the first in Europe" within this field and a possible solution to his brother's chronic constipation. See William James to Henry James, October 25 [1869], Skrupskelis and Berkeley, eds., *Correspondence*, 1:113.

128. James, "The Gospel," 119.

129. Ibid. Benedikt wrote about his concept of the "inner life" ("Binnenleben"), in two of his texts: Moriz Benedikt, "Second Life: Das Seelen-Binnenleben des gesunden und kranken Menschen: Vortrag für den internationalen medicinischen Congress in Rom (1894)," *Wiener Klinik*, no. 20 (1894); Moriz Benedikt, *Die Seelenkunde des Menschen als reine Erfahrungswissenschaft* (Leipzig: Reisland, 1895). The first, which was a small article published after he lectured on the topic in 1894, dealt only with one part of this alternative, inner life, namely, the "Seelen-Binnenleben" or "inner life of the soul." His subsequent book, however, included one section that comprised a reprint of his earlier article along with further discussion of a corporeal dimension to the inner life ("körperliches Binnenleben").

130. See, for example, Th. Ribot, *The Diseases of Personality*, 2nd ed. (Chicago: Open Court Publishing Company, 1895).

131. William James, "*As a Matter of Course, by Annie Payson Call* (1895)," in *Essays, Comments, and Reviews*, ed. Burkhardt, Bowers, and Skrupskelis, 503-4. See Annie Payson Call, *As a Matter of Course* (London: Samson Low & Co., 1895).

132. James, "The Gospel," 126-27.

133. Ibid., 128-29.

134. William James to Thomas Davidson, March 30, 1884, Skrupskelis and Berkeley, eds., *Correspondence*, 5:498.

135. Haley, *The Healthy Body*, 21.

136. Ray, *Mental Hygiene*, 139-40.

137. James, *Talks to Teachers*, 72-73. This passage also speaks to a theme that is discussed in chapter 5, "Politics and Pathology," namely, James's opposition to the idea that any one morbid psychological characteristic or deficiency was, in isolation, indicative of serious mental disease or incapacity for life.

138. Ibid., 54.

139. James, "The Gospel," 129.

140. James, "Draft and Outline," 48, 51, 55.

Chapter Three

1. James, *Pragmatism*, 131.

2. Perry wrote in his foreword that "James began his philosophical thinking about 1860, at a time when the armies of science and religion were being mobilized for the war which lasted out the century, and in which James sought to be a mediator." Later in his account of James's life he locates these warring factions inside James's adolescent psyche. He claims that James felt admiration for and an attraction toward science "on account of its fidelity to fact," while "at the same time he was haunted by a cosmic nostalgia—by those deeper doubts and hopes which are the perpetual spring of religion." He concludes that it was "inevitable, therefore, that [James] should look to philosophy for a way of reconciling science with the equally undeniable need for a sustaining faith." Perry, *Thought and Character*, 1:x, 450. Eugene Taylor embodies the two sides of James's "deep personal conflict" in two specific people whom he claims had a significant influence on James during the "seminal period" of his youth. In Taylor's account James's fa-

ther, Henry James Sr., personified "religion," and a family friend and positivist, Chauncey Wright, represented "science." He argues that James was buffeted between these two respected sages but finally managed to distance himself intellectually from them both and forge his own worldview, one that incorporated elements of their respective ideas. Taylor, *Consciousness*, 13.

3. William James to James John Garth Wilkinson, March 28, 1886, Skrupskelis and Berkeley, eds., *Correspondence*, 6:125; William James, "The Hidden Self," in *Essays in Psychology*, ed. Burkhardt, Bowers, and Skrupskelis, 248–49. In the folklore of many countries the seventh son of a seventh son was believed to have magical powers that included healing abilities.

4. William James, "*Rapport sur les progrès et la marche de la physiologie générale en France*, by Claude Bernard (1868)," in *Essays, Comments, and Reviews*, ed. Burkhardt, Bowers, and Skrupskelis, 225.

5. Ibid., 227.

6. Ibid., 228.

7. William James, "*Lectures on the Elements of Comparative Anatomy*, by Thomas Huxley (1865)," in *Essays, Comments, and Reviews*, ed. Burkhardt, Bowers, and Skrupskelis, 197.

8. William James to Thomas Wren Ward, March 1869, Skrupskelis and Berkeley, eds., *Correspondence*, 4:370.

9. William James, "What Psychical Research Has Accomplished (1896)," in *The Will to Believe*, ed. Burkhardt, Bowers, and Skrupskelis, 239.

10. James, "Lectures on the Elements," 202–3.

11. Ibid., 203.

12. According to Bruce Haley, it would appear that Huxley held a similar position. He recounts how, in his essay "On the Physical Basis of Life," Huxley called for the general adoption of a "materialistic terminology" in considering life, along with a "repudiation of materialism." Haley concludes that Huxley "felt that epistemologically it is more fruitful to consider life in its material aspect, even though we know it has a spiritual aspect as well. If we center our thoughts on those materialistic conceptions of life which are 'accessible' to us, we have a means of grasping the nonphysical world." Haley, *The Healthy Body*, 18–19.

13. C. Lloyd Tuckey, *Psycho-therapeutics, or Treatment by Hypnotism and Suggestion*, 2nd ed. (London: Baillière, Tindall, and Cox, 1890), 14. See also Henri Ellenberger, *The Discovery of the Unconscious: The History and Evolution of Dynamic Psychiatry* (New York: Basic Books, 1970), 85–87.

14. William James, "*Du sommeil et des états analogues* by Ambrose A. Liébeault (1868)," in *Essays, Comments, and Reviews*, ed. Burkhardt, Bowers, and Skrupskelis, 240. Although somnambulism, or "sleep-waking," occurred spontaneously in some subjects, Liébeault maintained that it could be artificially induced in most people.

15. Robert C. Fuller, *Mesmerism and the American Cure of Souls* (Philadelphia: University of Pennsylvania Press, 1982), 1–17.

16. In America, John Kearsley Mitchell, the father of the nineteenth-century neurologist Silas Weir Mitchell, wrote a series of essays that were subsequently edited and published by his son in 1859. One of these essays described his father's experiments with "animal magnetism." See S. Weir Mitchell, ed., *Five Essays by John Kearsley Mitchell M.D.* (Philadelphia: J. B. Lippincott & Co., 1859).

17. James, "Du sommeil," 242–45.

18. Ibid., 246.

19. William James to George Croom Robertson, August 13 [1885], Skrupskelis and Berkeley, eds., *Correspondence*, 6:63.

20. F. W. H. Myers, "On a Telepathic Explanation of Certain So-Called Spiritualistic Phenomena," *Proceedings of the Society for Psychical Research* 2 (1884): 234, 237.

21. William James to Frederic Myers, February 25, 1885, Skrupskelis and Berkeley, eds., *Correspondence*, 6:13.

22. In his subsequent 1887 essay on "The Consciousness of Lost Limbs," James wrote: "My final observations are on a matter which ought to interest students of 'psychic research.' Surely if there be any distant material object with which a man might be supposed to have clairvoyant or telepathic relations, that object ought to be his own cut-off arm or leg. Accordingly, a very wide-spread belief will have it that when the cut-off limb is maltreated in any way, the man, no matter where he is, will feel the injury. . . . But in none of the cases of my collection in which the writers seek to prove it does their conclusion inspire confidence." William James, "The Consciousness of Lost Limbs, 1887," in *Essays in Psychology*, ed. Burkhardt, Bowers, and Skrupskelis, 214.

23. William James, "Certain Phenomena of Trance," in *William James on Psychical Research*, ed. Gardner Murphy and Robert O. Ballou (London: Chatto and Windus, 1961), 104, 110.

24. James believed that this doctor "bears every appearance of being a fictitious being. His French . . . has been limited to a few phrases of salutation, . . . and the crumbs of information which he gives about his earthly career are, . . . few, vague and unlikely sounding." This did not imply that James doubted the veracity of the doctor's insights, however. Ibid., 105.

25. Richard Hodgson to Leonora Piper, October 20, 1891, cited in Alta L. Piper, *The Life and Work of Mrs. Piper* (London: Kegan Paul, Trench, Trubner & Co., Ltd., 1929), 76–77.

26. William James to Alice Howe Gibbens James [October 21, 1889]; William James to Alice Howe Gibbens James, October 21 [1889]; William James to Alice Howe Gibbens James, October 3, [1889], Skrupskelis and Berkeley, eds., *Correspondence*, 6:531, 540, 542.

27. Piper, *The Life*, 97.

28. Ibid., 16–17.

29. Frank Podmore, *Mesmerism and Christian Science: A Short History of Mental Healing* (London: Methuen & Co., 1909), 218, 249.

30. James John Garth Wilkinson, *The Greater Origins and Issues of Life and Death* (London: James Speirs, 1885).

31. W. M. Wilkinson, "The Cases of the Welsh Fasting Girl & Her Father: On the Possibility of Long-Continued Abstinence From Food" (London: J. Burns, 1870). This pamphlet is one item in a list, compiled in his wife's handwriting, of James's "valuable & much prized" manuscripts and books. See List of manuscripts and books prized by William James, William James Papers (MS Am 1092.9 [4581]), Houghton Library, Harvard University.

32. William James to James John Garth Wilkinson, March 28, 1886, Skrupskelis and Berkeley, eds., *Correspondence*, 6:124–25.

33. Ibid., 125.

34. James, *The Varieties*, 83.

35. For an account of some of the disagreements, see Eric Caplan, *Mind Games: American Culture and the Birth of Psychotherapy* (Berkeley: University of California Press, 2001), 77, 80.

36. The first of these periodicals, published in 1884, was *The Mind Cure and Science of Life Magazine*. See ibid. 76.

37. William James to Carl Stumpf, February 6, 1887; William James to Katharine James Prince, February 3, 1887, Skrupskelis and Berkeley, eds., *Correspondence*, 6:199, 204.

38. Shortly before James's visits to the "mind-cure doctress" he wrote to Dorr asking how negotiations on his behalf with a Mrs. Dresser were progressing. William James to George Bucknam Dorr, December 19, 1887, ibid., 578. Furthermore, the month after his visits to the mind-curist James wrote to a correspondent that Annetta Dresser is "inert" in his case. William James to Elizabeth Ellery Sedgwick, March 25, 1887, ibid., 582. See, in addition, James's letter to his wife, written a year later, which reveals that she had been wondering if a lightheartedness of hers could be due to a Mrs. Dresser. This suggests that she too had gone to her for treatment. William James to Alice Gibbens James, February 25, 1888, ibid., 329–30.

39. Charles S. Braden, *Spirits in Rebellion: The Rise and Development of New Thought* (Dallas: Southern Methodist University Press, 1970), 132, 160.

40. William James to Alice James, February 5, 1887, Skrupskelis and Berkeley, eds., *Correspondence*, 6:200–201.

41. William James to Katharine James Prince, February 3, 1887, ibid., 199.

42. Annetta Gertrude Dresser, *The Philosophy of P. P. Quimby: With Selections from His Manuscripts and a Sketch of His Life* (Boston: Geo. H. Ellis, 1895), 12, 16.

43. In contrast Warren Felt Evans, who was also treated by Quimby and went on to publicize his own interpretation of his healing methods, was an advocate of "magnetic healing." He wrote: "Magnetism . . . is identical with the universal life-principle, and is the medium through which our minds influence the minds of others, and a thought and will impulse are transmissible. To undertake to affect the mind of another without it is as absurd as to attempt to communicate sound through an absolute vacuum, or for a skillful mechanic to go forth to his work, but leave his tools behind." Warren Felt Evans, *Healing by Faith; or Primitive Mind-Cure, Elementary Lessons in Christian Philosophy and Transcendental Medicine* (London: Reeves and Turner, 1885), 164.

44. Dresser, *The Philosophy of P. P. Quimby*, 16, 40.

45. Ibid., 16, 68–70.

46. William James to Henry James, March 10, 1887, Ignas K. Skrupskelis and Elizabeth M. Berkeley, eds., *The Correspondence of William James*, vol. 2 (Charlottesville: University Press of Virginia, 1993), 59.

47. William James to Elizabeth Ellery Sedgwick Child, March 25, [1887], Skrupskelis and Berkeley, eds., *Correspondence*, 6:582.

48. In 1887 he sought treatment for insomnia, in 1893 for melancholy, in 1894 for insomnia and headache, in 1898 and 1899 he sought relief for unspecified symptoms, for nervous prostration in 1900, insomnia again in 1906, two separate courses of treatment for "anginoid pain" in 1907, and in 1909 he visited a Christian Scientist complaining of chest pains, respiratory problems, and nervous prostration. Emma Sutton, "Interpreting 'Mind-Cure': William James and the 'Chief Task . . . of the Science of Human Nature,'" *Journal of the History of the Behavioral Sciences* 48, no. 2 (2012): 125.

49. William James to Alice Howe Gibbens James, January 19, 1888, Skrupskelis and Berkeley, eds., *Correspondence*, 6:300.

50. William James to Alice Howe Gibbens James, April 9, 1888, ibid., 369. Annetta Dresser's son, Horatio, described this ideal mind-cure attitude in his later writings on the "New Thought": "We should anticipate good health, strength, happiness, an abundance of physical comfort, freshened faculties. . . . In a word we should be generally, definitely,

hourly, always optimistic, receptive to the good. Our wiser mental life should not, however end here, for we have the work to undo of all the gossips, meddlers, bad newsmongers, pessimists and trouble-breeders of the world. When we think of people, whether foes or friends, we are to hold only good thoughts concerning them. . . . We are to live in and for the good, think it, be it, do it, spread it abroad, invest our whole presence with its beauty and love." See Horatio Dresser, *Voices of Freedom and Studies in the Philosophy of Individuality* (New York: Knickerbocker Press, 1899), 40.

51. William James to Christine Ladd Franklin, April 12, 1888, Skrupskelis and Berkeley, eds., *Correspondence*, 6:374.

52. Pierre Janet, *L'Automatisme psychologique: Essai de psychologie expérimentale sur les formes inférieures de l'activité humaine* (Paris: Alcan, 1889).

53. William James to George Bucknam Dorr, October 25, 1889, William James Correspondence (MS Am 1092.1 [119]), Houghton Library, Harvard University.

54. For Janet's earlier writings, see Pierre Janet, "Note sur quelques phénomènes de somnambulisme," *Revue Philosophique* 21, no. 1 (1886); Pierre Janet, "Les Phases intermédiaires de l'hypnotisme," *Revue Scientifique*, 3rd ser., 2 (May 8 1886): 577–87; Pierre Janet, "Les Actes inconscients et le dédoublement de la personnalité pendant le somnambulisme provoqué," *Revue Philosophique* 22, no. 2 (1886); Pierre Janet, "Deuxième note sur le sommeil provoqué à distance et la suggestion mentale pendant l'état somnambulique," *Revue Philosophique* 22, no. 1 (1887); Pierre Janet, "L'Anesthésie systématisée et la dissociation des phénomènes psychologiques," *Revue Philosophique* 23, no. 1 (1887); Pierre Janet, "Les Actes inconscients et la mémoire pendant le somnambulisme," *Revue Philosophique* 25, no. 1 (1888). In the last of these articles Janet does mention, briefly, a young girl, "M.," who may be Marie, but he gives only scant details of her case and does not refer to her having been cured. See Janet, "Les actes inconscients et la mémoire."

55. James, "Hidden Self," 250, 258.

56. Ibid., 266.

57. Ibid.

58. Ibid., 266–67.

59. Alfred Binet, who later became the director of the Laboratory for Physiological Psychology at the Sorbonne, Paris, was another French investigator with an interest in the division of consciousness. He also mentioned novel practices, akin to Janet's, in his book entitled *Alterations of Personality*. There he discussed the possibility of taking a subject back to an earlier period of life, via suggestion, while they are in a trance state. The importance of such "retrospective suggestions," he stated, "has not yet been appreciated, . . . [and they] will certainly have, one day, important medical applications." He went on to explain that this practice "serves diagnosis by allowing us to see the origin and course of hysterical symptoms in all their details; and, on the other hand, it will perhaps be found that when the patient is carried back by this mental device to the time when the symptoms first appeared, he will be more amenable to suggestions of cure." Alfred Binet, *Alterations of Personality*, trans. Helen Green Baldwin (London: Chapman and Hall, Ltd., 1896), 268. Binet's text was first published in 1891, shortly after Janet's exposition of Marie's cure in his thesis. It is open to speculation as to why Binet did not cite his colleague's case study as an example, but relations between them were antagonistic. Despite popular assumptions, Freud's first publications on this theme postdate Janet's and Binet's, and James himself referred to Freud and Breuer's work as a "corroboration of Janet's views." William James, "'Ueber den psychischen Mechanismus hysterischer Phänomene,' by Josef Breuer and Sigmund Freud (1894)," in *Essays, Comments, and Reviews*, ed. Burkhardt, Bowers, and Skrupskelis, 474.

60. James, "Hidden Self," 265–66.

61. In his *Principles of Psychology* James referred any reader interested in the therapeutic applications of the hypnotic trance to a work by the German physician Albert Moll, which he described as "extraordinarily complete and judicious." The volume in question provides a good overview of the field of hypnotic healing around the time when Janet's thesis was published. See Albert Moll, *Hypnotism* (London: Walter Scott, 1890).

62. Tuckey, *Psycho-therapeutics*, 24.

63. In addition, Janet's association of *traumatic* life episodes with the onset of hysterical symptoms appears to have been derived from the work of his teacher, Charcot. Charcot had proposed that the nervous shock associated with traumatic events could create a state analogous to the hypnotic trance, thus facilitating the development of autosuggested ideas in the mind of the patient. See Ellenberger, *The Discovery of the Unconscious*, 91.

64. James further stated that this clarification of meaning owes everything to the concept of secondary levels of consciousness: "It seems to me a very great step to have ascertained that the secondary self, or selves, coexist with the primary one, the trance-personalities with the normal one, during the waking state." James, "Hidden Self," 267–68.

65. Daniel Hack Tuke, *Illustrations of the Influence of the Mind upon the Body in Health and Disease, Designed to Elucidate the Action of the Imagination*, 2nd ed. (Philadelphia: Henry C. Lea's Son & Co., 1884), 46.

66. There is evidence of a personal dimension to James's protestations that genuine disease is dismissed as arising from the patient's imagination. As a young man James defended the reality of his own illnesses to his brother Henry as follows: "I don't know whether you still consider my ailments to be imagination and humbug or not, but I know myself that they are as real as any one's ailments ever were." William James to Henry James, May 25, 1873, Skrupskelis and Berkeley, eds., *Correspondence*, 1:208.

67. In the manuscript notes for his 1896 Lowell Lectures on "Exceptional Mental States" James repeatedly emphasized this conclusion that hysteria is a "real disease, but *mental* disease." Hysterics genuinely suffered, he reiterated, "Shamming? No!"; "Consciousness splits . . . Not shammers then." William James, "Lectures on Abnormal Psychology (1895, 1896)," in *Manuscript Lectures*, ed. Burkhardt, Bowers, and Skrupskelis, 68–69.

68. William James, "The Medical Advertisement Abomination (1894)," in *Essays, Comments, and Reviews*, ed. Burkhardt, Bowers, and Skrupskelis, 143–44.

69. William James, "*Psychologie der Suggestion*, by Hans Schmidkunz (1892)," in *Essays, Comments, and Reviews*, ed. Burkhardt, Bowers, and Skrupskelis, 422–23.

70. Ibid., 423. This was also Myers's interpretation of the mechanism of "suggestion." See F. W. H. Myers, "The Subliminal Consciousness, Chapter 2: The Mechanism of Suggestion," *Proceedings of the Society for Psychical Research* 7 (1892).

71. James, "Hidden Self," 267.

72. The *Metaphysical Magazine* was a monthly publication devoted to the "study of the operations and phenomena of the human mind; and to a systematic inquiry into the faculties and functions . . . and attributes, of the spiritual man." "Announcement," *Metaphysical Magazine* 2 (July–December 1895): iv.

73. Leander Edmund Whipple, *The Philosophy of Mental Healing: A Practical Exposition of Natural Restorative Power* (New York: Metaphysical Publishing Company, 1893), 179.

74. William James, "The Philosophy of Mental Healing by Leander E. Whipple," in *Essays, Comments, and Reviews*, ed. Burkhardt, Bowers, and Skrupskelis, 476.

75. Myers coined the expression telepathy in an article in 1882: "We venture to introduce the words *Telaesthesia* and *Telepathy* to cover all cases of impression received at a distance without the normal operation of sense organs." F. W. H. Myers, "First Report of the Literary Committee," *Proceedings of the Society for Psychical Research* 1, no. 2 (1882): 147.

76. F. W. H. Myers, *Human Personality and Its Survival of Bodily Death* (London: Longmans, Green & Co., 1903), xxii.

77. F. W. H. Myers, "Note on a Suggested Mode of Psychical Interaction," in *Phantasms of the Living*, vol. 2, ed. Edmund Gurney, Frederic Myers, and Frank Podmore (Gainesville, FL: Scholars' Facsimiles & Reprints, 1970), 285.

78. F. W. H. Myers, "The Subliminal Consciousness, Chapter 1: General Characteristics of Subliminal Messages," *Proceedings of the Society for Psychical Research* 7 (1892): 306.

79. James, "Psychologie," 422.

80. James, "Hidden Self," 249.

81. Ibid., 247–48.

82. Ibid., 249.

83. Ibid. The other examples he used to prove his point were also of relevance to the medical profession. The "stigmatizations, invulnerabilities, instantaneous cures, inspired discourses, and demoniacal possessions, the records of which were shelved in our libraries but yesterday in the alcove headed 'Superstitions,' now, under the brand-new title of 'Cases of hystero-epilepsy,' are republished, reobserved, and reported with an even too credulous avidity." Ibid.

84. Ibid. By 1890 exhibitions of hypnotism as public entertainment had been prohibited by law in Switzerland and Holland. In addition, this topic was the subject of much debate at the international congress of physicians practicing hypnotism, which was held in Paris in 1889.

85. Ibid., 248–49.

86. William James, "On a Certain Blindness in Human Beings," in *Talks to Teachers*, ed. Burkhardt, Bowers, and Skrupskelis.

87. See James's letter to Wilkinson, in 1886, which was cited earlier: William James to James John Garth Wilkinson, March 28, 1886, Skrupskelis and Berkeley, eds., *Correspondence*, 6:124–25.

88. William James to George Croom Robertson, August 13 [1885], ibid., 62.

89. James, "Hidden Self," 250.

90. James, "Lectures on Abnormal," 63.

91. William James, "Address on the Medical Registration Bill (1898)," in *Essays, Comments, and Reviews*, ed. Burkhardt, Bowers, and Skrupskelis, 60.

92. William James, "Medical Registration Act (1894)," in *Essays, Comments, and Reviews*, ed. Burkhardt, Bowers, and Skrupskelis, 148; James, "Address on the Medical Registration Bill (1898)," 59–60.

93. James, "Address on the Medical Registration Bill (1898)," 59.

94. James, *The Varieties*, 102.

95. Ibid.

96. Ibid. Ramón del Castillo describes James's concept of religion in *The Varieties* as "individualistic" and "essentially *egotistic*" and argues that it owes much to the influence of his godfather, Ralph Waldo Emerson. See Ramón del Castillo, "The Glass Prison, Emerson, James and the Religion of the Individual," in *Fringes of Religious Experience: Cross Perspectives on William James's The Varieties of Religious Experience*, ed. Sergio Franzese and Felicitas Kraemer (Frankfurt: Ontos, 2007).

97. James, *The Varieties*, 104.

98. James, *Pragmatism*, 34.

99. James, *The Varieties*, 104–5.

100. Ibid., 105.

101. William James, "The Psychology of Belief," *Mind* 14, no. 55 (1889): 350.

102. William James, "*The Unseen Universe* by Peter Guthrie Tait and Balfour Stewart (1875)," in *Essays, Comments, and Reviews*, ed. Burkhardt, Bowers, and Skrupskelis, 293.

103. James, "Psychology of Belief," 351.

104. James, "What Psychical (1896)," 241.

105. Ibid.

106. William James, "What Psychical Research Has Accomplished (1892)," in *The Works of William James: Essays in Psychical Research*, ed. Frederick H. Burkhardt, Fredson Bowers, and Ignas K. Skrupskelis (Cambridge, MA: Harvard University Press, 1986), 105.

107. William James to Sarah Wyman Whitman, September 18, 1902, Skrupskelis and Berkeley, eds., *Correspondence*, 10:128.

108. William James to Rosina Hubley Emmet, August 25, 1902, ibid., 113.

Chapter Four

1. Sergio Franzese locates James's interest in the concept of energy in the context of contemporary scientific discussion, but he focuses on the physical, rather than the medical sciences. See Sergio Franzese, *The Ethics of Energy: William James's Moral Philosophy in Focus* (Frankfurt: Ontos, 2008).

2. For an overview of the medical significance of energy within late nineteenth- and early twentieth-century psychology, see Sonu Shamdasani, *Jung and the Making of Modern Psychology: The Dream of a Science* (Cambridge: Cambridge University Press, 2003), 202–6.

3. James, "Rationality, Activity and Faith," 81, 82, 84.

4. William James, "The Moral Philosopher and the Moral Life," in *The Will to Believe*, ed. Burkhardt, Bowers, and Skrupskelis, 161.

5. For James's description of God as an "infinite mind," see James, "The Dilemma of Determinism," 139.

6. William James to Thomas Wren Ward, January 7, 1868, Skrupskelis and Berkeley, eds., *Correspondence*, 4:250.

7. Ibid., 249.

8. William James to Thomas Davidson, January 8, 1882, Skrupskelis and Berkeley, eds., *Correspondence*, 5:195.

9. William James to Hannah Whitall Smith, May 11, 1886, Skrupskelis and Berkeley, eds., *Correspondence*, 6:138.

10. William James to Henry William Rankin, January 19, 1896, Skrupskelis and Berkeley, eds., *Correspondence*, 8:122.

11. Donald Duclow gives a brief account of this period of James's life as part of his discussion of the concept of the "sick soul" in *The Varieties*, but he does not explore the significance of these autobiographical circumstances as a context for the lectures as a whole. See Duclow, "William James, Mind-Cure and the Religion of Healthy-Mindedness," 51–52.

12. William James to Katharine Outram Rodgers, August 5, 1899, Skrupskelis and Berkeley, eds., *Correspondence*, 9:17.

13. William James to Eliza Putnam Webb Gibbens, August 2, 1899, ibid., 11–12.

14. William James to Henry James III, August 14, 1899; William James to Francis Boot, October 6, 1899, ibid., 24, 57.

15. William James to William Wilberforce Baldwin, November 2, 1899; William James to François Pillon, November 26, 1899; William James to James Jackson Putnam, December 2, 1899; William James to Wincenty Lutoslawski, January 1, 1900, ibid., 72, 89–90, 92, 114.

16. The Gifford Lectures were given at the Scottish universities and funded at the behest of Adam Lord Gifford. Their remit was to promote and spread the study of natural theology, meaning theological knowledge that is supported by observation of the natural world. For more details about the history of the lectures, see Hendrika Vande Kemp, "The Gifford Lectures on Natural Theology: Historical Background to James's Varieties," *Streams of William James* 4, no. 3 (2002).

17. William James to Francis Rollins Morse, December 23, 1899, Skrupskelis and Berkeley, eds., *Correspondence*, 9:105.

18. William James to Théodore Flournoy, January 1, 1900, ibid., 113.

19. William James, "Appendix 4: Notebook Containing Titles and Outlines," in *The Varieties*, ed. Burkhardt, Bowers, and Skrupskelis, 498–99.

20. William James to Dickinson Sergeant Miller, January 18, 1900, Skrupskelis and Berkeley, eds., *Correspondence*, 9:129.

21. There has been a considerable amount of scholarship dealing with *The Varieties of Religious Experience*. For alternative readings of the text and its significance, see Michael Ferrari, ed., "The Varieties of Religious Experience: Centenary Essays," special issue, *Journal of Consciousness Studies* 9, nos. 9–10 (2002); Charles Taylor, *Varieties of Religion Today: William James Revisited* (Cambridge, MA: Harvard University Press, 2002); Paul Jerome Croce and John Snarey, eds., *Streams of William James* 4, no. 3 (2002); John Snarey and Paul Jerome Croce, eds., *Streams of William James* 5, no. 3 (2003); Wayne Proudfoot, ed., *William James and a Science of Religions: Reexperiencing The Varieties of Religious Experience* (New York: Columbia University Press, 2004); Jeremy Carrette, ed., *William James and The Varieties of Religious Experience: A Centenary Celebration* (London: Routledge, 2005); Franzese and Kraemer, eds., *Fringes of Religious Experience*. See also the introductory essays by Eugene Taylor and Jeremy Carrette in William James, *The Varieties of Religious Experience: Centenary Edition*, ed. Eugene Taylor and Jeremy Carrette (London: Routledge 2002).

22. William James to Thomas Davidson, February 16, 1900, Skrupskelis and Berkeley, eds., *Correspondence*, 9:143–44.

23. James, "Appendix 4," 500.

24. Ibid.

25. Ibid., 501.

26. Ibid.

27. James, *The Varieties*, 45–46.

28. Ibid., 46, 48.

29. Ibid., 114, 121–22, 136.

30. Ibid., 115.

31. Ibid., 119.

32. Ibid., 122–24, 127.

33. Ibid., 131, 138.

34. Ibid., 140–41.

35. Smith Baker, "Etiological Significance of Heterogeneous Personality," *Journal of Nervous and Mental Disease* 18 (1893): 672–73.

36. Baker stressed the importance of a good education in limiting the pathological consequences of a heterogeneous personality: "it will be of incalculable worth to so educate and reinforce the better, healthier characteristics that they shall become permanently dominant in the personal life. The relations of pedagogy to the prophylaxis of disease, thus become of [significant] importance with those of heterogeneous personality." Ibid., 674.

37. James, *The Varieties*, 142.

38. Ibid., 146.

39. Ibid., 160.

40. Ibid., 161.

41. Ibid., 162.

42. Ibid., 161.

43. In James's *Principles of Psychology*, the will is also the locus of potential pathology. In his chapter on "Will" he stated that *"unhealthiness of will may . . . come about in many ways. . . . If we compare the outward symptoms of perversity together, they fall into two groups. . . . Briefly, we may call them respectively the obstructed and the explosive will."* He described the obstructed will as a case in which "impulsion is insufficient or inhibition in excess." He did not explore the concept of the divided will in this context, however, which supports the hypothesis (to be discussed below) that James only developed this specific psychopathological model after he published *The Principles*, during the mid-1890s. James, *The Principles*, 2:1143–44, 1152.

44. See, for example, Bleuler's introductory chapter to Jung's *Studies in Word-Association*, which was first published in German in 1906. There Bleuler wrote: "Every psychical activity rests upon the interchange of the material derived from sensation and from memory traces. . . . The will, despite all attempts, seems almost to elude psychological investigation. A man's psyche can be fully described without making use of this vague concept. Psychopathology frequently ignores the will altogether." Eugen Bleuler, "Upon the Significance of Association Experiments," in *Studies in Word-Association: Experiments in the Diagnosis of Psychopathological Conditions Carried Out at the Psychiatric Clinic of the University of Zurich*, ed. C. G. Jung (New York: Moffat, Yard & Company, 1919), 1.

45. James, *The Varieties*, 188.

46. The timing of their publications supports this hypothesis. James first described how a clash of scientific and religious loyalties, or in other words a divided self, can lead to severe melancholy in his address "Is Life Worth Living?," which he delivered in 1895. James had been in contact with Baker since 1891, and Baker's paper on heterogeneous personality was published in 1893. James appears to have supplemented Baker's conceptual framework with some elements derived from Benedikt's notion of the "Binnenleben," which Benedikt first spoke about in public a year later, in 1894. The influence of both men's work is evident in "Is Life Worth Living?," and Benedikt's work was explicitly cited by James, albeit anonymously. William James, "Is Life Worth Living?," in *The Will to Believe*, ed. Burkhardt, Bowers, and Skrupskelis.

47. James, *The Varieties*, 170.

48. Edwin Diller Starbuck, *The Psychology of Religion: An Empirical Study of the Growth of Religious Consciousness* (London: Walter Scott, Ltd., 1899). William Woody locates James's description of conversion with respect to the work of Starbuck and other contemporary psychologists, such as G. Stanley Hall, who were interested in religious phenomena. See William Douglas Woody, "Varieties of Religious Conversion: William James in Historical and Contemporary Contexts," *Streams of William James* 5, no. 1 (2003).

49. James, *The Varieties*, 171. James implied that Starbuck's focus on the importance of self-surrender, or the giving up of the personal will, was his pupil's own novel contribution to the psychology of conversion experiences.

50. Ibid., 173.

51. Ibid., 173–74.

52. There has, to date, been a reluctance to locate James's accounts of religious experience in a medical context, perhaps because the extent of James's medical interests and concerns has not previously been appreciated. For example, Eugene Taylor discusses James's "depth psychology"; however, he does not read James's interest in the subconscious primarily in terms of its practical, medical implications, but rather as part of James's "study of mystical consciousness" and "transcendent experience," which, he argues, were integral to his ideas about a "growth-oriented dimension to personality." Eugene Taylor, "William James and Depth Psychology," in "The Varieties of Religious Experience: Centenary Essays," special issue, *Journal of Consciousness Studies* 9, nos. 9–10 (2002) 16, 32. In his article, psychotherapist Mark Krejci discusses the "connections [he has] made between James's descriptions of individual religious experience and the experiences of [his] clients as they dealt with religious issues during therapy." He prefaces his study, however, with the statement that, in *The Varieties*, "James's goal was to explore personal religious experience, not to discuss the therapeutic ramifications of this experience." Mark J. Krejci, "Self and Healthy-Mindedness: James's *Varieties* and Religion in Psychotherapy," *Streams of William James* 5, no. 1 (2003): 16.

53. James, *The Varieties*, 96–97.

54. Starbuck, *The Psychology of Religion*, 149–51.

55. James, *The Varieties*, 165, 169.

56. William James to Henry Rankin, February 1, 1897, Skrupskelis and Berkeley, eds., *Correspondence*, 8:227–28.

57. James, *The Varieties*, 403. Anne Taves makes this point about the subconscious in her essay about James's use of Janet and Myers's ideas in *The Varieties*. See Ann Taves, "The Fragmentation of Consciousness and *The Varieties of Religious Experience*: William James's Contribution to a Theory of Religion," in *William James and a Science of Religions: Reexperiencing The Varieties of Religious Experience*, ed. Wayne Proudfoot (New York: Columbia University Press, 2004).

58. James, *The Varieties*, 197.

59. William James, "Review of *Human Personality and Its Survival of Bodily Death*, by Frederic W. H. Myers (1903)," in *William James on Psychical*, ed. Murphy and Ballou, 230.

60. Ibid.

61. William James, "The Powers of Men," in *The Works of William James: Essays in Religion and Morality*, ed. Frederick H. Burkhardt, Fredson Bowers, and Ignas K. Skrupskelis (Cambridge, MA: Harvard University Press, 1982), 152.

62. James, *The Varieties*, 196.

63. Ibid., 408.

64. Michael Ferrari mentions James's depiction of the energizing properties of religious faith, but he does not place this Jamesian characterization in a medical context. See Michael Ferrari, "The Personal Paradox of William James's *Varieties*," *Streams of William James* 4, no. 3 (2002): 20–21.

65. Myers hypothesized that during a state of hypnosis contact is made with the subliminal consciousness and its "recuperative energies." He wrote: "what we have in effect been doing with the aid of these hypnotic artifices is simply to energise Life. . . . Typi-

cal of Life is its self-adaptive power, its capacity . . . of righting the organism when it has been in any way injured; — that *vis medicatrix Naturae* which is the inmost secret of the living organism. Hypnotism has shown us this *vis medicatrix* in an unprecedentedly definite and controllable form." Myers, *Human Personality*, 216.

66. James, *The Varieties*, 44.

67. Ibid., 397–99.

68. Ibid., 407–8.

69. Ibid., 410.

70. Ibid., 114.

71. Ibid., 410, 412.

72. William James, *A Pluralistic Universe*, ed. Frederick H. Burkhardt, Fredson Bowers, and Ignas K. Skrupskelis, The Works of William James (Cambridge, MA: Harvard University Press, 1977), 138.

73. Ibid., 139–40.

74. James, *The Varieties*, 168-9.

75. William James to Alice Runnells James, December 15, 1909, Ignas K. Skrupskelis and Elizabeth M. Berkeley, eds., *The Correspondence of William James*, vol. 12 (Charlottesville: University of Virginia Press, 2004), 387.

76. William James, "The Energies of Men," in *Essays in Religion and Morality*, ed. Burkhardt, Bowers, and Skrupskelis, 129–30.

77. Ibid., 130.

78. Pierre Janet, *Les obsessions et la psychasthénie* (Paris: Félix Alcan, 1903).

79. James, "The Energies of Men," 130.

80. See William James to Wincenty Lutoslawski, October 31, 1902; William James to Giulio Cesare Ferrari, September 1, 1902, Skrupskelis and Berkeley, eds., *Correspondence*, 10:116, 118.

81. James, "The Energies of Men," 131.

82. Ibid., 132–33.

83. James, *Talks to Teachers*, 41–42; James, "Is Life Worth Living?," 45.

84. James, *The Varieties*, 213–15.

85. Henry James, ed., *The Letters of William James*, vol. 2 (London: Longmans, Green, and Co., 1920), 284.

86. William James, "The Moral Equivalent of War," in *Essays in Religion and Morality*, ed. Burkhardt, Bowers, and Skrupskelis, 164–73.

87. William James to Henry James III, August 14, 1899, Skrupskelis and Berkeley, eds., *Correspondence*, 9:23.

88. Mitchell, *Wear and Tear*, 7–8.

89. James, "The Moral Equivalent," 169–71.

90. James, *The Varieties*, 291.

91. James, "The Energies of Men," 134–36.

92. Ibid.

93. Ibid., 136–38.

94. William James to Wincenty Lutoslawski, May 6, 1906, Ignas K. Skrupskelis and Elizabeth M. Berkeley, eds., *The Correspondence of William James*, vol. 11 (Charlottesville: University of Virginia Press, 2003), 220.

95. James, "The Energies of Men," 139.

96. Ibid.

97. Ibid., 145.

98. Ibid.

99. William James to Ferdinand Canning Scott Schiller, April 7, 1906, Skrupskelis and Berkeley, eds., *Correspondence*, 11:197.

100. John R. Shook and Hugh P. McDonald, eds., *F. C. S. Schiller on Pragmatism and Humanism: Selected Writings, 1891–1939* (New York: Humanity Books, 2008), 12.

101. William James, "G. Papini and the Pragmatist Movement in Italy," *Journal of Philosophy, Psychology and Scientific Methods* 3, no. 13 (1906): 339.

102. Ibid., 340.

103. For a detailed analysis of James's understanding of the relationship between an organism and its environment and his critiques of Spencer's ideas, see chapter 2 of Pearce, *Pragmatism's Evolution*, 58–100.

104. In Perry's words, the works of Spencer became James's favored "source books for the illustration of philosophical error—'almost a museum of blundering reason.'" Perry, *Thought and Character*, 1:475. For James's criticisms of Spencer's work, see William James, "Great Men and Their Environment," in *The Will to Believe*, ed. Burkhardt, Bowers, and Skrupskelis; William James, "Some Remarks on Spencer's Definition of Mind," in *The Works of William James: Essays in Philosophy*, ed. Frederick H. Burkhardt, Fredson Bowers, and Ignas K. Skrupskelis (Cambridge, MA: Harvard University Press, 1978). In the latter essay, published in 1878, James deconstructed the normative content of Spencer's ideas about mental evolution and the environment. At this stage, however, he left intact his implicit definition of health or "physical well-being" as the ability to "adjust" or conform to the environment.

105. William James to Wincenty Lutoslawski, May 6, 1906, Skrupskelis and Berkeley, eds., *Correspondence*, 11:221.

Chapter Five

1. See, for example, Deborah J. Coon, "One Moment in the World's Salvation: Anarchism and the Radicalization of William James," *Journal of American History* 83, no. 1 (1996); Robert B. Westbrook, "Lewis Mumford, John Dewey, and the 'Pragmatic Acquiescence,'" in *Lewis Mumford: Public Intellectual*, ed. Thomas P. Hughes and Agatha C. Hughes (New York: Oxford University Press, 1990); Livingston, *Pragmatism and the Political Economy*; Joshua I. Miller, *Democratic Temperament: The Legacy of William James* (Lawrence: University Press of Kansas, 1997); Brian Lloyd, *Left Out: Pragmatism, Exceptionalism, and the Poverty of American Marxism, 1890–1920* (Baltimore: Johns Hopkins University Press, 1997). For a comprehensive survey of the literature on James's political affiliations and her own interpretation of James's stance toward individualism and capitalism, see Bordogna, *William James at the Boundaries*, 189–217.

2. William James to Ernest Howard Crosby, October 23, 1901, Skrupskelis and Berkeley, eds., *Correspondence*, 9:551.

3. William James, "What Makes a Life Significant," in *Talks to Teachers*, ed. Burkhardt, Bowers, and Skrupskelis, 165–66.

4. Micah Hester examines how James's definition of a meaningful life may be deployed within contemporary health-care contexts in D. Micah Hester, *End-of-Life Care and Pragmatic Decision Making: A Bioethical Perspective* (Cambridge: Cambridge University Press, 2009).

5. These prominent voices included, not least, those responsible for drafting state legislation in America from the middle decades of the nineteenth century onward who chose, in increasing numbers, to disenfranchise those deemed to be "idiots," "insane per-

sons," or "under guardianship." Alexander Keyssar, *The Right to Vote: The Contested History of Democracy in the United States* (New York: Basic Books, 2009), 288.

6. In this way, my analysis of James's politics comprises a contribution to the field of disability studies. Catherine Kudlick, in her review essay "Disability History: Why We Need Another 'Other,'" argues that "in terms of raw numbers and lived experience, [disability] occupies a place comparable to gender and race in defining the human condition." She makes the case that the social category of disability "should sit squarely at the center of historical inquiry, both as a subject worth studying in its own right and as one that will provide scholars with a new analytic tool for exploring power itself." Catherine Kudlick, "Disability History: Why We Need Another 'Other,'" *American Historical Review* 108, no. 3 (2003): 765, 767.

7. William James to Charles Renouvier, May 8, 1882, Skrupskelis and Berkeley, eds., *Correspondence*, 5:209.

8. William James to Francis Herbert Bradley, July 9, 1895, Skrupskelis and Berkeley, eds., *Correspondence*, 8:52. For discussion of the links between health and work in James's ideas, see chapter 1, "Misery and Metaphysics," and chapter 2, "Health and Hygiene."

9. See chapter 1, "Misery and Metaphysics," for an account of this stage in James's life and ethical thought.

10. William James to Frederic Myers, December 17, 1893; William James to Théodore Flournoy, December 31, 1893, Ignas K. Skrupskelis and Elizabeth M. Berkeley, eds., *The Correspondence of William James*, vol. 7 (Charlottesville: University of Virginia Press, 1999), 474, 479.

11. William James to Henry Lee Higginson, March 26, 1892; William James to Hugo Münsterberg, May 15, 1892; William James to Hugo Münsterberg, July 6, 1893; William James to Frederic Myers, July 16, [1893]; William James to Dickinson Sergeant Miller, November 19, 1893, ibid., 253, 267, 434, 439, 469.

12. William James to Shadworth Hollway Hodgson, March 28, 1892, ibid., 255. James's son, Henry James III, wrote that his father "called psychology 'a nasty little subject,' according to Professor Palmer, and added, 'all one cares to know lies outside.'" James, ed. *The Letters*, 2:2.

13. William James to George Holmes Howison, October 28, [1893]; William James to Théodore Flournoy, August 1894, Skrupskelis and Berkeley, eds., *Correspondence*, 7: 466, 536.

14. William James to Théodore Flournoy, December 31, 1893, ibid., 479.

15. For a survey of the earlier formulations of degeneration theory by the French alienists Bénédict Morel and Valentin Magnan, see Dowbiggin, *Inheriting Madness*; Ian Dowbiggin, "Back to the Future: Valentin Magnan, French Psychiatry, and the Classification of Mental Diseases, 1885–1925," *Social History of Medicine* 9, no. 3 (1996). According to Dowbiggin the popularity of the theory of mental degeneracy peaked during the 1890s. Dowbiggin, *Inheriting Madness*, 162. For other historical treatments of the notion of degeneration, see Sander L. Gilman and J. Edward Chamberlin, eds., *Degeneration: The Dark Side of Progress* (New York: Columbia University Press, 1985); Daniel Pick, *Faces of Degeneration: A European Disorder, c. 1848–c. 1918* (Cambridge: Cambridge University Press, 1989).

16. Joseph Collins, "*Degeneration* by Max Nordau," *Journal of Nervous and Mental Disease* 20, no. 7 (1895): 453.

17. The predilection among the members of such groups for forming societies is seen, by Nordau, as symptomatic of their underlying degenerate natures: "When authors or artists consciously and intentionally meet together and found an aesthetic school . . . as a

rule it is disease. . . . Under the influence of [a pathological] obsession, a degenerate mind promulgates some doctrine or other. . . . Other degenerate, hysterical, neurasthenical minds flock around him, receive from his lips the new doctrine, and live thenceforth only to propagate it." Max Nordau, *Degeneration* (London: William Heinemann, 1895), 29–31.

18. Ibid., 43.

19. Ibid., 537.

20. Ibid., 536. In contrast, other authors writing about degeneracy, such as the Italian physician Cesare Lombroso, associated it with a criminal underclass.

21. According to James's 1895 review of Jules Dallemagne's treatise on Degeneration, *Dégénérés et déséquilibrés*, Dallemagne also embraced hysteria and neurasthenia within his concept of the "'degenerative' type." William James, *"Dégénérés et déséquilibrés, by Jules Dallemagne (1895),"* in *Essays, Comments, and Reviews*, ed. Burkhardt, Bowers, and Skrupskelis, 505.

22. Collins, *"Degeneration by Max Nordau,"* 454.

23. Roy Porter, "Nervousness, Eighteenth and Nineteenth Century Style: From Luxury to Labour," in *Cultures of Neurasthenia from Beard to the First World War*, ed. Marijke Gijswijt-Hofstra and Roy Porter (New York: Rodopi B.V., 2001), 39; William James to Hugo Munsterberg, July 6, 1893, Skrupskelis and Berkeley, eds., *Correspondence*, 7:434.

24. Volker Roelcke, "Electrified Nerves, Degenerated Bodies: Medical Discourses on Neurasthenia in Germany, circa 1880–1914," in *Cultures of Neurasthenia*, ed. Gijswijt-Hofstra and Porter, 181.

25. Nordau, *Degeneration*, 557.

26. Laura Goering, "'Russian Nervousness': Neurasthenia and National Identity in Nineteenth Century Russia," *Medical History* 47, no. 1 (2003): 31. Other historians, such as Joost Vijselaar, have contributed nuanced accounts of how the concept of neurasthenia was put to use in different European countries. He points out that in the Netherlands, for example, neurasthenic discourses never acquired the pessimistic and strongly biological overtones that characterized the French and German writings. Joost Vijselaar, "Neurasthenia in the Netherlands," in *Cultures of Neurasthenia*, ed. Gijswijt-Hofstra and Porter, 239–43.

27. William James, *"Genie und Entartung, by William Hirsch (1895),"* in *Essays, Comments, and Reviews*, ed. Burkhardt, Bowers, and Skrupskelis, 513.

28. William James to Alfred Marston Allen, October 26, 1882, Skrupskelis and Berkeley, eds., *Correspondence*, 5:284.

29. There was already evidence of a potential family weakness in this area, as James's brother Bob was believed to suffer from severe mental disease. In 1888 James confided to his wife that Bob's recent behavior made him fear that his brother's "brain-degeneracy [was] advancing." William James to Alice Howe Gibbens James, February 5, 1888, Skrupskelis and Berkeley, eds., *Correspondence*, 6:308.

30. Susan E. Gunter, *Alice in Jamesland: The Story of Alice Howe Gibbens James* (Lincoln: University of Nebraska Press, 2009), 210.

31. William James, *"Degeneration, by Max Nordau,"* in *Essays, Comments, and Reviews*, ed. Burkhardt, Bowers, and Skrupskelis, 508.

32. Nordau, *Degeneration*, 537, 556–57.

33. James, *"Degeneration,"* 513.

34. Nordau first published the original German edition of his book *Entartung* in two volumes in 1892 and 1893, while James was traveling around Europe with his family. Max Nordau, *Entartung*, 2 vols. (Berlin: Duncker, 1892–93).

35. A letter sent by James to Augustus Lowell about his proposed series of lectures at

the Lowell Institute lists ten topics, but these were subsequently compressed into only eight. William James to Augustus Lowell, April 18, 1896, Skrupskelis and Berkeley, eds., *Correspondence*, 8:141–42.

36. James, "Lectures on Abnormal," 63.

37. In the notes for his lectures on "Exceptional Mental States" this section is in note form, but James made the same point in his *Principles of Psychology*. In the chapter on "Will" he described how an "insistent idea" can take over certain patients' lives when "what would be for most people the passing suggestion of a possibility becomes a gnawing, craving urgency to act." He added, however, that "most people have the potentiality of this disease. To few has it not happened to conceive, after getting into bed, that they may have forgotten to lock the front door, or to turn out the entry gas. And few of us have not on some occasion got up to repeat the performance, less because they believed in the reality of its omission than because only so could they banish the worrying doubt and get to sleep." James, *The Principles*, 2:1148, 1152.

38. James, "Lectures on Abnormal," 55.

39. Georges Canguilhem, *On the Normal and the Pathological*, trans. Carolyn R. Fawcett (Dordrecht: D. Reidel Publishing Co., 1978), 13, 15.

40. Dickinson Sergeant Miller to Henry James III, August 24, 1917, William James Papers (MS Am 1092.10 [122]), Houghton Library, Harvard University.

41. Charles W. Eliot, *Public Opinion and Sex Hygiene: An Address Delivered at the Fourth International Congress on School Hygiene, at Buffalo, New York, August 27th, 1913* (New York: American Federation for Sex Hygiene, 1913), 12–13.

42. James, "Lectures on Abnormal," 82.

43. James, "Dégénérés et déséquilibrés"; William James, "*Entartung und Genie*, by Cesare Lombroso (1895)," in *Essays, Comments, and Reviews*, ed. Burkhardt, Bowers, and Skrupskelis; James, "Degeneration"; James, "Genie und Entartung."

44. William James to Henry Holt, April 20, 1895, Skrupskelis and Berkeley, eds., *Correspondence*, 8:27.

45. This review was written by the American neurologist Smith Ely Jelliffe. Jelliffe, "*Genius and Degeneration* by Dr. William Hirsch," *Journal of Nervous and Mental Diseases* 24, no. 6 (1897).

46. William Hirsch, *Genius and Degeneration: A Psychological Study*, trans. from the 2nd German ed. (London: William Heinemann, 1897), 198, 330.

47. This was the adjective that Jelliffe used to characterize the book in his review. Jelliffe, "*Genius and Degeneration.*"

48. James, "Genie und Entartung," 509. James was referring to Cesare Lombroso and J. F. Nisbet who also wrote on degeneration theory. Lombroso was particularly concerned with the criminological aspect of degeneration, namely, the inheritance of criminal traits and accompanying physical characteristics or "stigmata." James cited Nisbet's book, J. F. Nisbet, *The Insanity of Genius and the General Inequality of Human Faculty Physiologically Considered*, 3rd ed. (London: Ward & Downey, 1893), in James, *The Varieties*.

49. James, "Genie und Entartung," 512–13.

50. Ibid., 513.

51. James, "Lectures on Abnormal," 82.

52. Ibid., 83. James referred to the lectures as having been shaped "towards optimistic and hygienic conclusions." See William James to George Holmes Howison, April 5, 1897, Skrupskelis and Berkeley, eds., *Correspondence*, 8:256. As the notes do not seem to indicate that James spent any time discussing the topic of habit formation or any other

typically hygienic practices, it would appear that he believed that dispelling fears about mental pathology constituted, in and of itself, a hygienic step. This would accord with sentiments expressed by James in his *Talks to Teachers* and his essay on "The Gospel of Relaxation," which were discussed in chapter 2, "Health and Hygiene." It is interesting to note that James included the same exhortation at the end of his last lecture of the "Exceptional Mental States" series, his set of book reviews on degeneration, and his address "Is Life Worth Living?" On each occasion his instruction essentially comprised the phrase, "Be not afraid of life." Reading these rallying cries in conjunction with each other it becomes apparent that they are all speaking to a specific subtext. It is not life in general that James perceived as potentially frightening, but a life menaced by one's own morbid mind. In the review article on degeneration his message is most clearly articulated. There James ended with the words "we should broaden our notion of [mental] health instead of narrowing it; . . . we should regard no single element of weakness as fatal—in short . . . we should *not be afraid of life.*" James, "Genie und Entartung," 513.

53. Alongside these elements of continuity, James's philosophical understandings of religion and God also underwent significant transformations. See chapter 3, "Religion and Regeneration," and chapter 4, "Energy and Endurance."

54. Perry, *Thought and Character*, 2:208.

55. James, *The Varieties*, 296. Gregory Moore discusses Nietzsche's familiarity with the degeneration literature and how themes drawn from these texts both "inflect and infect so many of Nietzsche's own writings." Gregory Moore, "Nietzsche, Degeneration, and the Critique of Christianity," *Journal of Nietzsche Studies* 19 (Spring 2000): 2.

56. James, *The Varieties*, 296. See Friedrich Nietzsche, *On the Genealogy of Morality*, ed. Keith Ansell-Pearson, trans. Carol Diethe (Cambridge: Cambridge University Press, 1994), 95.

57. James, *The Varieties*, 297.

58. James referred to "the Francises, Bernards, Luthers, Loyolas, Wesleys, Channings, Moodys, Gratrys, the Phillips Brookes, the Agnes Joneses, Margaret Hallahans, and Dora Pattisons." Ibid., 299.

59. Thus, set in the context of the discussion that follows below, they were the ultimate Jamesian spiritual heroes, leading the fight against the evil of illness in their public and personal lives. In his previous chapter, "Saintliness," James noted that "the nursing of the sick is a function to which the religious seem strongly drawn, even apart from the fact that church traditions set that way." Ibid., 229.

60. Kim Townsend, *Manhood*, 17, 97–103. Although his two youngest brothers enlisted, James did not fight in the Civil War. Louis Menand explores James's decision not to sign up in Menand, *Metaphysical Club*, 73–77.

61. Maria H. Frawley, *Invalidism and Identity in Nineteenth-Century Britain* (Chicago: University of Chicago Press, 2004), 34.

62. Susan Coolidge, *What Katy Did* (Boston: Roberts Brothers, 1872).

63. James, *The Varieties*, 299.

64. William James to Alice Smith, April 17, 1900, Skrupskelis and Berkeley, eds., *Correspondence*, 9:190.

65. James, *The Varieties*, 299.

66. The English biologist Thomas Huxley discussed the phrase "survival of the fittest" and challenged its contemporary application to human society on the same grounds. See Thomas H. Huxley, *Evolution and Ethics* (London and New York: Macmillan and Co., 1893), 32–33.

67. James, *The Varieties*, 294, 298, 299.

68. James, "Is Life Worth Living?," 48. George Cotkin devotes a chapter to James's "Discourse of Heroism" in his book *William James: Public Philosopher* but does not discuss James's linking of heroism with the suffering of the invalid. Cotkin, *William James*.

69. James, "Is Life Worth Living?," 48.

70. Ibid., 52.

71. Jeremy Carrette discusses James's construction of a pet perspective in *Pragmatism* and *A Pluralistic Universe* in Carrette, *Hidden Religious Imagination*, xv.

72. James, "Is Life Worth Living?," 53.

73. William James to Robertson James, February 20 [1875], Skrupskelis and Berkeley, eds., *Correspondence*, 4:508–9.

74. William James to Katharine James Prince, July 11, 1886, Skrupskelis and Berkeley, eds., *Correspondence*, 6:150.

75. James, "Is Life Worth Living?," 47, 56; James, "The Moral Philosopher," 160–61; James, "The Moral Equivalent," 172. Kim Townsend offers an alternative reading of James's battlefield rhetoric in Townsend, *Manhood at Harvard*, 169.

76. James, *The Varieties*, 9, 289.

77. Ibid., 289.

78. Ibid.

79. Charles Binet-Sanglé, *La folie de Jésus: Son hérédité, sa constitution, sa physiologie* (Paris: A. Maloine, 1908).

80. James, *The Varieties*, 20.

81. Henry Maudsley, *Natural Causes and Supernatural Seemings*, 3rd ed. (London: Kegan Paul, Trench, Trübner & Co., 1897), 277–78.

82. James, *The Varieties*, 21.

83. James made a similar argument in his 1899 essay "On a Certain Blindness in Human Beings." There he discussed how "the whole of truth . . . is revealed to [no] single observer" though each makes a contribution. Making implicit reference to the frequent associating of the criminal and the diseased within the degeneration literature, he insisted that "even prisons and sick-rooms have their special revelations." William James, "On a Certain Blindness," 149.

84. Dickinson Sergeant Miller to William James, February 26, 1899, Skrupskelis and Berkeley, eds., *Correspondence*, 8:501. Regarding James's specific linking of the "psychopathic temperament" with mystical insight, Miller's observation may be true. In a more general sense, however, as noted earlier, the historical research of Maria Frawley indicates that many of his nineteenth-century contemporaries understood there to be an association between ill health and spiritual experiences. She asserts that "serious or protracted illness was widely believed to promote conversion, the doctrine most central to evangelical theology. Evangelical beliefs in the spiritual transformation that followed conversion also dovetailed with beliefs about bodily and psychic transformations within the sickroom." Frawley, *Invalidism and Identity*, 34.

85. James, *The Varieties*, 28.

86. Benjamin Paul Blood to William James, June 11, 1887; William James to Benjamin Paul Blood, June 19, 1887, Skrupskelis and Berkeley, eds., *Correspondence*, 6:230, 232, 234.

87. William James to John Jay Chapman, January 31, 1898, Skrupskelis and Berkeley, eds., *Correspondence*, 8:341.

88. James, *A Pluralistic Universe*, 69.

89. James, *The Varieties*, 230.

90. William James, "Introduction to *The Literary Remains of the Late Henry James*," in *Essays in Religion and Morality*, ed. Burkhardt, Bowers, and Skrupskelis, 30–37.

91. James, *The Varieties*, 112.

92. Ibid., 136.

93. This Jamesian link between evil and illness is discussed at length in chapter 1, "Misery and Metaphysics."

94. James, *The Varieties*, 122. James made a similar observation in the notes for his Lowell lectures. There he stated, "Melancholy! Gives truer values." James, "Lectures on Abnormal," 63.

95. Burkhardt, Bowers, and Skrupskelis, eds., *Essays, Comments, and Reviews*, 567nn.

96. Ibid., 556–57nn.

97. James, ed., *The Letters*, 2:67, 68.

98. William James to James Jackson Putnam, March 2, 1898, Skrupskelis and Berkeley, eds., *Correspondence*, 8:348.

99. William James to John Jay Chapman, March 4, 1898, ibid., 350. The Dreyfus affair, as it came to be known, began when Dreyfus was arrested for treason in 1894, and his plight led to fierce public debate and accusations of anti-Semitism.

100. In his address James acknowledged that his protest would affect his standing in the eyes of his medical colleagues but stated "my duty is to the larger society, the Commonwealth. I cannot look on passively; and I must urge my point." James, "Address on the Medical Registration Bill (1898)," 58.

101. William James to James Jackson Putnam, March 10, 1898, Skrupskelis and Berkeley, eds., *Correspondence*, 8:351.

102. James, "Address on the Medical Registration Bill (1898)," 62. In light of the connection James made between the abuses of medical and religious authority, it is interesting to refer to his account, in *The Varieties*, of "religion's wicked practical partner" and "religion's wicked intellectual partner." The former he labels "the spirit of corporate dominion," and the latter he describes as "the spirit of dogmatic dominion" or "the passion for laying down the law in the form of an absolutely closed-in theoretic system." In *The Varieties* James took pains to point out that such innate "tribal instincts" are not a necessary part of religion per se, but often parade themselves under the cloak of piety. Similarly, it would seem that for James the admirable ideals of healing are also often accompanied by baser human instincts, namely, the same spirits of corporate and dogmatic dominion that led, in the name of religion, to "The baiting of Jews, . . . the stoning of Quakers and ducking of Methodists." James, *The Varieties*, 271.

103. James, "Address on the Medical Registration Bill (1898)," 60–61.

104. James, "Medical Registration Act (1894)," 147–48.

105. Ibid.

106. James, "Address on the Medical Registration Bill (1898)," 59.

107. James publicly called for asylum reform as early as 1875, in one of his many book reviews on morbid psychology. William James, "The Borderlands of Insanity, by Andrew Wynter (1875)," in *Essays, Comments, and Reviews*, ed. Burkhardt, Bowers, and Skrupskelis, 315–16.

108. William James to Alice Howe Gibbens James, May 26 [1890], Skrupskelis and Berkeley, eds., *Correspondence*, 7:41. Edward Cowles was the superintendent of the McLean Asylum at Somerville and an acquaintance of James.

109. Clifford Whittingham Beers, *A Mind That Found Itself: An Autobiography* (London, Bombay, Calcutta, and Madras: Longmans, Green and Co., 1917), 88.

110. William James to Clifford Whittingham Beers, July 1, 1906, William James Correspondence (MS Am 1092.1 [9]), Houghton Library, Harvard University.

111. Norman Dain, *Clifford W. Beers: Advocate for the Insane* (Pittsburgh: University

of Pittsburgh Press, 1980), 22, 74–75. James also sent a copy to an Italian correspondent, Dr. Ferrari, who was head of the provincial asylum of Bologna and editor of the "leading Italian psychological review." Ferrari was impressed by Beers's book and wrote a favorable notice of it. Later he translated it for an Italian readership. William James to Clifford Whittingham Beers, February 9, 1909, William James Correspondence (MS Am 1092.1 [9]), Houghton Library, Harvard University.

112. James's son Henry James III testified that his father had cared deeply about Beers's campaign. "He even departed, in its interest, from his fixed policy of 'keeping out of Committees and Societies.' He lived long enough to know that the movement had begun to gather momentum; and he drew great satisfaction from the knowledge." James, ed., *The Letters*, 2:273.

113. "I am going yo [*sic*] tackle ex-president Eliot who is on the general Education Board, about the possibility of getting a grant say of 10–20,000 for this year only." William James to Clifford Whittingham Beers, March 7, 1910, William James Correspondence (MS Am 1092.1 [9]), Houghton Library, Harvard University.

114. Rockefeller was the father-in-law of the American philosopher and psychologist Charles Strong, a graduate of Harvard during James's tenure there and someone with whom James remained in close personal contact. James met Rockefeller for the first time while visiting Strong in 1903 and described him as a "most complex old scoundrel." William James to William James Jr., January 26, 1903, Skrupskelis and Berkeley, eds., *The Correspondence*, 10:185.

115. Dain, *Clifford Beers*, 128.

116. Ibid., 169–70, 187.

117. Clifford Whittingham Beers to William James, January 16, 1910, Skrupskelis and Berkeley, eds., *The Correspondence*, 12:415.

118. William James to Clifford Whittingham Beers, January 17, 1910, ibid., 416–17.

119. William James to Clifford Whittingham Beers, April 21, 1907, William James Correspondence (MS Am 1092.1 [9]), Houghton Library, Harvard University.

120. Whittingham Beers, *A Mind*, 1.

121. James, *The Varieties*, 192.

122. Whittingham Beers, *A Mind*, 115.

123. William James to Clifford Whittingham Beers, April 15, 1908, William James Correspondence (MS Am 1092.1 [9]), Houghton Library, Harvard University.

124. Draft of William James to John Davison Rockefeller, January 31, 1909, William James Papers (MS Am 1092.9 [3553]), Houghton Library, Harvard University.

Conclusion

1. William James, Answers to J. B. Pratt's questionnaire on religion, William James Papers (MS Am 1092.9 [4474]), Houghton Library, Harvard University.

2. James, Miscellaneous notes, William James Papers.

3. Ibid.

4. Bjork, *The Centre of His Vision*, 261.

5. William James to Alice Howe Gibbens James, March 4 [1888], Skrupskelis and Berkeley, eds., *Correspondence*, 6:338–39.

6. James, "Appendix 1," 185; James, *Talks to Teachers*, 27.

7. James, *The Varieties*, 297, 299.

8. James, "What Psychical (1892)," 241.

9. James, "G. Papini," 340.

10. William James to Wincenty Lutoslawski, May 6, 1906, Skrupskelis and Berkeley, eds., *Correspondence*, 11:221.

11. William James to Sarah Wyman Whitman, September 18, 1902, Skrupskelis and Berkeley, eds., *Correspondence*, 10:128.

12. James Jackson Putnam, "William James," *Atlantic Monthly*, December 1910, 839.

13. James, "The Gospel," 118.

14. James, "Draft and Outline," 51.

15. James, "The Energies of Men," 130–31.

16. Ibid., 143.

17. Ibid.

18. William James to Sarah Wyman Whitman, September 18, 1902, Skrupskelis and Berkeley, eds., *Correspondence*, 10:128.

19. Putnam, "William James," 835.

20. William James to Alice Howe Gibbens James, July 26, 1896, Skrupskelis and Berkeley, eds., *Correspondence*, 8:173.

21. Marion Hamilton Carter to William James, June 16, 1903, Skrupskelis and Berkeley, eds., *Correspondence*, 10:268–69.

22. Charles T. Sempers to Henry James III, July 15, 1920, William James Papers (MS Am 1092.10 [155]), Houghton Library, Harvard University.

23. Sarah Harvey Porter to Henry James III, November 6, 1911, William James Papers (Ms Am 1092.10 [137]), Houghton Library, Harvard University.

24. Henry M. Rogers to Henry James III, November 28, 1911, William James Papers (MS Am 1092.10 [146]), Houghton Library, Harvard University.

ARCHIVAL SOURCES

Alice Howe Gibbens James and Henry James III, carte-de-visite (Boston, ca. 1881; photographer: Allen & Rowell), James family additional papers (MS Am 2955 [51], Box: 1). Houghton Library, Harvard University.

Answers to J. B. Pratt's questionnaire on religion, William James Papers (MS Am 1092.9 [4474]). Houghton Library, Harvard University.

Cadaver heads and caricature of a man (ca. 1864–69), William James drawings (MS Am 1092.2 [44], Box: 1). Houghton Library, Harvard University.

Charles T. Sempers to Henry James III, July 15, 1920, William James Papers (MS Am 1092.10 [155]). Houghton Library, Harvard University.

Diary 1 (1868–73), William James Papers (MS Am 1092.9 [4550]). Houghton Library, Harvard University, accessed July 13, 2022, https://hollisarchives.lib.harvard.edu /repositories/24/digital_objects/14160.

Dickinson Sergeant Miller to Henry James III, August 24, 1917, William James Papers (MS Am 1092.10 [122]). Houghton Library, Harvard University.

Diplomas, degrees, notifications of appointments, etc., William James Papers (MS Am 1092.9–1092.12, MS Am 1092.9 [4571], Box: 40). Houghton Library, Harvard University.

Draft of William James to John Davison Rockefeller, January 31, 1909, William James Papers (MS Am 1092.9 [3553]). Houghton Library, Harvard University.

Henry James Sr. to Henry James, March 18, 1873, William James Papers (MS Am 1092.9 [4199]). Houghton Library, Harvard University.

Henry M. Rogers to Henry James III, November 28, 1911, William James Papers (MS Am 1092.10 [146]). Houghton Library, Harvard University.

List of manuscripts and books prized by William James, William James Papers (MS Am 1092.9 [4581]). Houghton Library, Harvard University.

Miscellaneous notes, William James Papers (MS Am 1092.9 [4473]). Houghton Library, Harvard University.

Notebook 26, William James Papers (MS Am 1092.9 [4520]). Houghton Library, Harvard University.

Sarah Harvey Porter to Henry James III, November 6, 1911, William James Papers (Ms Am 1092.10 [137]). Houghton Library, Harvard University.

William James in Boston with book, portrait photograph (ca. 1869; photographer: J. B.

Hawes), Letters to William James from various correspondents and photograph album (MS Am 1092 [1185], 12, Box: 12). Houghton Library, Harvard University.

William James sitting with Mrs. Walden in séance (undated; photographer: Miss Carter),
Letters to William James from various correspondents and photograph album (MS
Am 1092 [1185], 1, Box: 12). Houghton Library, Harvard University.

William James to Clifford Whittingham Beers, July 1, 1906, William James Correspondence (MS Am 1092.1 [9]). Houghton Library, Harvard University.

William James to Clifford Whittingham Beers, April 21, 1907, William James Correspondence (MS Am 1092.1 [9]). Houghton Library, Harvard University.

William James to Clifford Whittingham Beers, April 15, 1908, William James Correspondence (MS Am 1092.1 [9]). Houghton Library, Harvard University.

William James to Clifford Whittingham Beers, February 9, 1909, William James Correspondence (MS Am 1092.1 [9]). Houghton Library, Harvard University.

William James to Clifford Whittingham Beers, March 7, 1910, William James Correspondence (MS Am 1092.1 [9]). Houghton Library, Harvard University.

William James to George Bucknam Dorr, October 25, 1889, William James Correspondence (MS Am 1092.1 [119]). Houghton Library, Harvard University.

William James with daughter, Margaret Mary James (1892, photographer: May Whit-
well), Letters to William James from various correspondents and photograph album
(MS Am 1092 [1185], 3, Box: 12). Houghton Library, Harvard University.

BIBLIOGRAPHY

Allen, Gay Wilson. *William James: A Biography.* New York: Viking Press, 1967.

Andrews, Jonathan. "Winslow, Forbes Benignus (1810–1874)." In *Oxford Dictionary of National Biography.* Oxford: Oxford University Press, 2004.

"Announcement." *Metaphysical Magazine* 2 (July–December 1895).

Aurelius, Marcus. *The Meditations of Marcus Aurelius Antoninus.* Translated by A. S. L. Farquharson. Edited by R. B. Rutherford. Oxford: Oxford University Press, 1989.

Baker, Smith. "Etiological Significance of Heterogeneous Personality." *Journal of Nervous and Mental Disease* 18 (1893): 664–74.

Bartlett, Elisha. *Obedience to the Laws of Health, a Moral Duty: A Lecture Delivered before the American Physiological Society, January 30, 1838.* Boston: Julius A. Noble, 1838.

Beard, George M. *A Practical Treatise on Nervous Exhaustion (Neurasthenia): Its Symptoms, Nature, Sequences, Treatment.* New York: William Wood & Company, 1880.

Bending, Lucy. *The Representation of Bodily Pain in Late Nineteenth-Century English Culture.* Oxford: Oxford University Press, 2000.

Benedikt, Moriz. *Die Seelenkunde des Menschen als reine Erfahrungswissenschaft.* Leipzig: Reisland, 1895.

Benedikt, Moriz. "Second Life: Das Seelen-Binnenleben des gesunden und kranken Menschen: Vortrag für den internationalen medicinischen Congress in Rom (1894)." *Wiener Klinik,* no. 20 (1894): 127–38.

Bentham, Jeremy. *An Introduction to the Principles of Morals and Legislation.* Edited by J. H. Burns and H. L. A. Hart. Oxford: Clarendon Press, 1996.

Berkley Fletcher, Holly. *Gender and the American Temperance Movement of the Nineteenth Century.* London: Routledge, 2007.

"Bertrand on Suicide." *American Journal of Insanity* 14 (1857): 207–14.

Bertrand, Louis. *Traité du suicide: Considéré dans ses rapports avec la Philosophie, la Théologie, la Médecine et la Jurisprudence.* Paris: J. B. Baillière, 1857.

Bezly Thorne, W. *Schott Methods of the Treatment of Chronic Diseases of the Heart with an Account of the Nauheim Baths, and of the Therapeutic Exercises.* London: J. & A. Churchill, 1896.

Binet, Alfred. *Alterations of Personality.* Translated by Helen Green Baldwin. London: Chapman and Hall, Ltd., 1896.

Binet-Sanglé, Charles. *La folie de Jésus: Son hérédité, sa constitution, sa physiologie.* Paris: A. Maloine, 1908.

Bird, Graham H. "Moral Philosophy and the Development of Morality." In *The Cambridge Companion to William James.* Cambridge: Cambridge University Press, 1997.

Bjork, Daniel W. *William James: The Centre of His Vision.* New York: Columbia University Press, 1988.

Blandford, George Fielding. "Prevention of Insanity (Prophylaxis)." In *A Dictionary of Psychological Medicine: Giving the Definition, Etymology, and Synonyms of the Terms Used in Medical Psychology with the Symptoms, Treatment, and Pathology of Insanity and the Law of Lunacy in Great Britain and Ireland,* edited by D. Hack Tuke. London: J. & A. Churchill, 1892.

Bleuler, Eugen. "Upon the Significance of Association Experiments." In *Studies in Word-Association: Experiments in the Diagnosis of Psychopathological Conditions Carried Out at the Psychiatric Clinic of the University of Zurich,* edited by C. G. Jung. New York: Moffat, Yard & Company, 1919.

Boddice, Rob, ed. *Pain and Emotion in Modern History.* Basingstoke: Palgrave Macmillan, 2014.

Boddice, Rob. *The Science of Sympathy: Morality, Evolution and Victorian Civilization.* Urbana: University of Illinois Press, 2016.

Bordogna, Francesca. *William James at the Boundaries: Philosophy, Science and the Geography of Knowledge.* Chicago: University of Chicago Press, 2008.

Bourke, Joanna. *The Story of Pain: From Prayer to Painkillers.* Oxford: Oxford University Press, 2014.

Bowditch, Henry Ingersoll. *Public Hygiene in America: Being the Centennial Discourse Delivered before the International Medical Congress, Philadelphia, September, 1876.* Boston: Little, Brown and Company, 1877.

Braden, Charles S. *Spirits in Rebellion: The Rise and Development of New Thought.* Dallas: Southern Methodist University Press, 1970.

Bunge, G. "Die Alkoholfrage, Ein Vortrag." Basel: L. Reinhardt, 1894.

Burkhardt, Frederick H., Fredson Bowers, and Ignas K. Skrupskelis, eds. *The Works of William James: Essays, Comments, and Reviews.* Cambridge, MA: Harvard University Press, 1987.

Burkhardt, Frederick H., Fredson Bowers, and Ignas K. Skrupskelis, eds. *The Works of William James: Essays in Psychology.* Cambridge, MA: Harvard University Press, 1983.

Burkhardt, Frederick H., Fredson Bowers, and Ignas K. Skrupskelis, eds. *The Works of William James: Manuscript Lectures.* Cambridge, MA: Harvard University Press, 1988.

Burkhardt, Frederick H., Fredson Bowers, and Ignas K. Skrupskelis, eds. *The Works of William James: Talks to Teachers on Psychology and to Students on Some of Life's Ideals.* Cambridge, MA: Harvard University Press, 1983.

Bynum, William F. "Chronic Alcoholism in the First Half of the 19th Century." *Bulletin of the History of Medicine* 42, no. 2 (1968): 160–85.

Call, Annie Payson. *As a Matter of Course.* London: Samson Low & Co., 1895.

Call, Annie Payson. *Power through Repose.* London: Gay and Hancock, 1914.

Canguilhem, Georges. *On the Normal and the Pathological.* Translated by Carolyn R. Fawcett. Dordrecht: D. Reidel Publishing Co., 1978.

Caplan, Eric. *Mind Games: American Culture and the Birth of Psychotherapy.* Berkeley: University of California Press, 2001.

Capps, Donald. "Relaxed Bodies, Emancipated Minds, and Dominant Calm." *Journal of Religion and Health* 48, no. 3 (2009): 368–80.

Carpenter, William B. *Principles of Mental Physiology, with Their Applications to the Train-ing and Discipline of the Mind, and the Study of Its Morbid Conditions.* New York: D. Appleton & Company, 1874.

Carpenter, William B. *Principles of Mental Physiology, with Their Applications to the Train-ing and Discipline of the Mind, and the Study of Its Morbid Conditions.* 4th (1876) ed. London: Routledge/Thoemmes Press, 1993.

Carrette, Jeremy. "Growing Up Zig-Zag: Reassessing the Transatlantic Legacy of William James." In *William James and the Transatlantic Conversation: Pragmatism, Pluralism, and Philosophy of Religion,* edited by Martin Halliwell and Joel D. S. Rasmussen. Ox-ford: Oxford University Press, 2014.

Carrette, Jeremy, ed. *William James and The Varieties of Religious Experience: A Centenary Celebration.* London: Routledge, 2005.

Carrette, Jeremy. *William James's Hidden Religious Imagination: A Universe of Relations.* London and New York: Routledge, 2013.

Chapman, John Jay. "William James: A Portrait." *Harvard Graduates' Magazine,* 1910, 233–37.

Clark, Elizabeth B. "'The Sacred Rights of the Weak': Pain, Sympathy, and the Culture of Individual Rights in Antebellum America." *Journal of American History* 82, no. 2 (1995): 463–93.

Clouston, T. S. *Clinical Lectures on Mental Diseases.* London: J. & A. Churchill, 1883.

Clouston, T. S. *Female Education from a Medical Point of View.* Edinburgh: Macniven & Wallace, 1882.

Collins, Joseph. "*Degeneration* by Max Nordau." *Journal of Nervous and Mental Disease* 20, no. 7 (1895): 453–55.

"Confessions of an English Opium-Eater 1." *London Magazine* 4, no. 21 (1821): 293–312.

"Confessions of an English Opium-Eater 2." *London Magazine* 4, no. 22 (1821): 353–79.

Coolidge, Susan. *What Katy Did.* Boston: Roberts Brothers, 1872.

Coon, Deborah. "One Moment in the World's Salvation: Anarchism and the Radicaliza-tion of William James." *Journal of American History* 83, no. 1 (1996): 70–99.

Cotkin, George. *William James: Public Philosopher.* Baltimore: John Hopkins Univer-sity Press, 1990.

Croce, Paul J. "James and Medicine: Reckoning with Experience." In *The Oxford Hand-book of William James,* edited by Alexander Klein. Accessed May 10, 2021, https://www.oxfordhandbooks.com/view/10.1093/oxfordhb/9780199395699.001.0001/oxfordhb-9780199395699-e-29.

Croce, Paul Jerome. *Science and Religion in the Era of William James.* Vol. 1, *Eclipse of Cer-tainty, 1820–1880.* Chapel Hill and London: University of North Carolina Press, 1995.

Croce, Paul J. *Young William James Thinking.* Baltimore: John Hopkins University Press, 2018.

Croce, Paul Jerome, and John Snarey, eds. *Streams of William James* 4, no. 3 (2002).

Dain, Norman. *Clifford W. Beers: Advocate for the Insane.* Pittsburgh: University of Pitts-burgh Press, 1980.

del Castillo, Ramón "The Glass Prison, Emerson, James and the Religion of the Indi-vidual." In *Fringes of Religious Experience: Cross Perspectives on William James's The Varieties of Religious Experience,* edited by Sergio Franzese and Felicitas Kraemer. Frankfurt: Ontos, 2007.

De Quincey, Thomas. *The Note Book of an English Opium-Eater.* Boston: Ticknor and Fields, 1855.

Deutsch, Albert. "The History of Mental Hygiene." In *One Hundred Years of American*

Psychiatry, edited by J. K. Hall, Gregory Zilboorg, and Henry Alden Bunker. New York: Columbia University Press, 1944.

Dixon, Thomas. *From Passions to Emotions: The Creation of Secular Psychological Category*. Cambridge: Cambridge University Press, 2003.

Dixon, Thomas. *The Invention of Altruism: Making Moral Meanings in Victorian Britain*. Oxford: Oxford University Press/British Academy, 2008.

Dods, Marcus, ed. *The Works of Aurelius Augustine, Bishop of Hippo*. Vol. 9, *On Christian Doctrine; the Enchiridion; on Catechising; and on Faith and the Creed*. Edinburgh: T. & T. Clark, 1873.

Donne, John. *Biathanatos: A Declaration of that Paradoxe, or Thesis, that Selfe-homicide is not so naturally Sinne, that it may never be Otherwise*. 1647.

Donne, John. *Biathanatos: A Modern-Spelling Critical Edition*. Edited by Michael Rudick and M. Pabst Battin. New York: Garland Publishing Company, 1982.

Dowbiggin, Ian. "Back to the Future: Valentin Magnan, French Psychiatry, and the Classification of Mental Diseases, 1885–1925." *Social History of Medicine* 9, no. 3 (1996): 383–408.

Dowbiggin, Ian R. *Inheriting Madness: Professionalisation and Psychiatric Knowledge in Nineteenth Century France*. Berkeley: University of California Press, 1991.

Dresser, Annetta Gertrude. *The Philosophy of P. P. Quimby: With Selections from His Manuscripts and a Sketch of His Life*. Boston: Geo. H. Ellis, 1895.

Dresser, Horatio. *Voices of Freedom and Studies in the Philosophy of Individuality*. New York: Knickerbocker Press, 1899.

Duclow, Donald F. "William James, Mind-Cure and the Religion of Healthy-Mindedness." *Journal of Religion and Health* 41, no. 1 (2002): 45–56.

Dumas, Jean. *Traité du suicide, ou, Du meurtre volontaire de soi-même*. Amsterdam: D. J. Changuion, 1773.

Dylan Knapp, Krister. *William James: Psychical Research and the Challenge of Modernity*. Chapel Hill: University of North Carolina Press, 2017.

Eigen, Joel. *Witnessing Insanity: Madness and Mad-Doctors in the English Court*. New Haven, CT: Yale University Press, 1995.

Eliot, Charles. "Introduction." In Herbert Spencer, *Essays on Education and Kindred Subjects*, edited by Charles Eliot. New York: E. P. Dutton & Co., 1911.

Eliot, Charles W. *Public Opinion and Sex Hygiene: An Address Delivered at the Fourth International Congress on School Hygiene, at Buffalo, New York, August 27th, 1913*. New York: American Federation for Sex Hygiene, 1913.

Ellenberger, Henri. *The Discovery of the Unconscious: The History and Evolution of Dynamic Psychiatry*. New York: Basic Books, 1970.

Ellsworth, Phoebe C. "William James and Emotion: Is a Century Worth of Fame Worth a Century of Misunderstanding?" *Psychological Review* 101, no. 2 (1994): 222–29.

Ernst, Waltraud, ed. *Histories of the Normal and the Abnormal: Social and Cultural Histories of Norms and Normativity*. London and New York: Routledge, 2006.

Étoc-Demanzy, Gustave-François. *Recherches statistiques sur le suicide, appliques à l'hygiene publique et à la medicine legale*. Paris: Germer-Bailliere, 1844.

Evans, Warren Felt. *Healing by Faith; or Primitive Mind-Cure, Elementary Lessons in Christian Philosophy and Transcendental Medicine*. London: Reeves and Turner, 1885.

Feinstein, Howard M. *Becoming William James*. Ithaca, NY: Cornell University Press, 1999.

Feinstein, Howard M. "The Use and Abuse of Illness in the James Family Circle: A View of Neurasthenia as a Social Phenomenon." In *Our Selves/Our Past: Psychological Ap-*

proaches to American History, edited by Robert J. Brugger. Baltimore: John Hopkins University Press, 1981.

Ferrari, Michael. "The Personal Paradox of William James's *Varieties.*" *Streams of William James* 4, no. 3 (2002): 13–22.

Ferrari, Michael, ed. "The Varieties of Religious Experience: Centenary Essays." Special issue, *Journal of Consciousness Studies* 9, no. 9–10 (2002).

Feuchtersleben, Baron Ernst von. *Hygiene of the Mind (Zur Diatetik der Seele).* Translated by F. C. Sumner. 3rd (1910) ed. New York: Macmillan Co., 1933.

Feuchtersleben, Ernest von. *The Dietetics of the Soul.* 7th ed. New York: C. S. Francis & Co., 1854.

Fiske, John. "Chauncey Wright." In *The Evolutionary Philosophy of Chauncey Wright*, vol. 3, *Influence and Legacy*, edited by Frank X. Ryan. Bristol: Thoemmes Press, 2000.

Forel, August. "Die Trinklitten, ihre hygienische und sociale Bedeutung. Ihre Beziehungen zur Akademischen Jugend." Basel: L. Reinhardt, 1894.

Foster, Grace. "The Psychotherapy of William James." *Psychoanalytic Review* 32 (1945): 300–318.

Foust, Mathew A. "William James and the Promise of *Pragmatism.*" *William James Studies* 2 (2007). Accessed July 1, 2021, https://williamjamesstudies.org/william-james-and-the-promise-of-pragmatism/.

Franzese, Sergio. *The Ethics of Energy: William James's Moral Philosophy in Focus.* Frankfurt: Ontos, 2008.

Franzese, Sergio, and Felicitas Kraemer, eds. *Fringes of Religious Experience: Cross-Perspectives on William James's The Varieties of Religious Experience.* Frankfurt: Ontos, 2007.

Frawley, Maria H. *Invalidism and Identity in Nineteenth-Century Britain.* Chicago: University of Chicago Press, 2004.

Fuller, Robert C. *Mesmerism and the American Cure of Souls.* Philadelphia: University of Pennsylvania Press, 1982.

Gilman, Sander L., and J. Edward Chamberlin, eds. *Degeneration: The Dark Side of Progress.* New York: Columbia University Press, 1985.

Goering, Laura. "'Russian Nervousness': Neurasthenia and National Identity in Nineteenth Century Russia." *Medical History* 47, no. 1 (2003): 23–46.

Goldstein, Jan. *Console and Classify: The French Psychiatric Profession in the Nineteenth Century.* Chicago: University of Chicago Press, 1987.

Gunter, Susan E. *Alice in Jamesland: The Story of Alice Howe Gibbens James.* Lincoln: University of Nebraska Press, 2009.

Gusfield, Joseph. *Symbolic Crusade: Status Politics and the American Temperance Movement.* Urbana and London: University of Illinois Press, 1963.

Hale, Sir Matthew. *The History of the Pleas of the Crown*, vol. 2. Edited by Sollom Emlyn. London, 1736.

Haley, Bruce. *The Healthy Body and Victorian Culture.* Cambridge, MA: Harvard University Press, 1978.

Hall, Lesley. "'It Was Affecting the Medical Profession': The History of Masturbatory Insanity Revisited." *Paedagogica Historica* 39, no. 6 (2003): 685–99.

Halliwell, Martin. "Morbid and Positive Thinking: William James, Psychology and Illness." In *William James and the Transatlantic Conversation: Pragmatism, Pluralism, and Philosophy of Religion*, edited by Martin Halliwell and Joel D. S. Rasmussen. Oxford: Oxford University Press, 2014.

Halliwell, Martin, and Joel D. S. Rasmussen, eds. *William James and the Transatlan-*

tic Conversation: Pragmatism, Pluralism, and Philosophy of Religion. Oxford: Oxford University Press, 2014.

Herbert, Wm. W. "The Forcible Feeding of the Insane." *British Medical Journal* 1, no. 1731 (March 1894): 462.

Hester, D. Micah. *End-of-Life Care and Pragmatic Decision Making: A Bioethical Perspective.* Cambridge: Cambridge University Press, 2009.

Hickman, Timothy A. "'Mania Americana': Narcotic Addiction and Modernity in the United States, 1870–1920." *Journal of American History* 90, no. 4 (2004): 1269–94.

Hinton, James, ed. *Physiology for Practical Use: By Various Writers.* 2 vols. London: Henry S. King & Co., 1874.

Hirsch, William. *Genius and Degeneration: A Psychological Study.* Translated from the 2nd German ed. London: William Heinemann, 1897.

Hoblyn, Richard D. *A Dictionary of Terms Used in Medicine and the Collateral Sciences.* 9th ed. London: Whittaker & Co., 1868.

Hookway, Christopher. "Logical Principles and Philosophical Attitudes: Pierce's Response to James's Pragmatism." In *The Cambridge Companion to William James*, edited by Ruth Anna Putnam. Cambridge: Cambridge University Press, 1997.

Hume, David. *Essays on Suicide and the Immortality of the Soul.* London, 1783.

Huxley, Thomas H. *Evolution and Ethics.* London and New York: Macmillan and Co., 1893.

James, Henry, ed. *The Letters of William James.* 2 vols. London: Longmans, Green, and Co., 1920.

James, Henry. "The Nature of Evil." In *Henry James Senior: A Selection of His Writings*, edited by Giles Gunn. Chicago: American Library Association, 1974.

James, William. "Address on the Medical Registration Bill (1898)." In *The Works of William James: Essays, Comments, and Reviews*, edited by Frederick H. Burkhardt, Fredson Bowers, and Ignas K. Skrupskelis. Cambridge, MA: Harvard University Press, 1987.

James, William. "Appendix 1: Notes and Draft for Talks to Teachers." In *The Works of William James: Talks to Teachers on Psychology and to Students on Some of Life's Ideals*, edited by Frederick H. Burkhardt, Fredson Bowers, and Ignas K. Skrupskelis. Cambridge, MA: Harvard University Press, 1983.

James, William. "Appendix 4: Notebook Containing Titles and Outlines." In *The Works of William James: The Varieties of Religious Experience*, edited by Frederick H. Burkhardt, Fredson Bowers, and Ignas K. Skrupskelis. Cambridge, MA: Harvard University Press, 1985.

James, William. "Are We Automata?" In *The Works of William James: Essays in Psychology*, edited by Frederick H. Burkhardt, Fredson Bowers, and Ignas K. Skrupskelis. Cambridge, MA: Harvard University Press, 1983.

James, William. "*As a Matter of Course, by Annie Payson Call* (1895)." In *The Works of William James: Essays, Comments, and Reviews*, edited by Frederick H. Burkhardt, Fredson Bowers, and Ignas K. Skrupskelis. Cambridge, MA: Harvard University Press, 1987.

James, William. "The Borderlands of Insanity, by Andrew Wynter (1875)." In *The Works of William James: Essays, Comments, and Reviews*, edited by Frederick H. Burkhardt, Fredson Bowers, and Ignas K. Skrupskelis. Cambridge, MA: Harvard University Press, 1987.

James, William. "Certain Phenomena of Trance." In *William James on Psychical Research*, edited by Gardner Murphy and Robert O. Ballou. London: Chatto and Windus, 1961.

James, William. "Chauncey Wright." In *The Evolutionary Philosophy of Chauncey Wright*, vol. 3, *Influence and Legacy*, edited by Frank X. Ryan. Bristol: Thoemmes Press, 2000.

James, William. "The Consciousness of Lost Limbs, 1887." In *The Works of William James: Essays in Psychology*, edited by Frederick H. Burkhardt, Fredson Bowers, and Ignas K. Skrupskelis. Cambridge, MA: Harvard University Press, 1983.

James, William. "Contributions to Mental Pathology, by Isaac Ray (1873)." In *The Works of William James: Essays, Comments, and Reviews*, edited by Frederick H. Burkhardt, Fredson Bowers, and Ignas K. Skrupskelis. Cambridge, MA: Harvard University Press, 1987.

James, William. "*Degeneration*, by Max Nordau (1895)." In *The Works of William James: Essays, Comments, and Reviews*, edited by Frederick H. Burkhardt, Fredson Bowers, and Ignas K. Skrupskelis. Cambridge, MA: Harvard University Press, 1987.

James, William. "*Dégénérés et déséquilibrés*, by Jules Dallemagne (1895)." In *The Works of William James: Essays, Comments, and Reviews*, edited by Frederick H. Burkhardt, Fredson Bowers, and Ignas K. Skrupskelis. Cambridge, MA: Harvard University Press, 1987.

James, William. "The Dilemma of Determinism." In *The Works of William James: The Will to Believe and Other Essays in Popular Philosophy*, edited by Frederick H. Burkhardt, Fredson Bowers, and Ignas K. Skrupskelis. Cambridge, MA: Harvard University Press, 1979.

James, William. "Draft and Outline of a Lecture on 'The Effects of Alcohol' (1895)." In *The Works of William James: Manuscript Lectures*, edited by Frederick H. Burkhardt, Fredson Bowers, and Ignas K. Skrupskelis. Cambridge, MA: Harvard University Press, 1988.

James, William. "Draft on Brain Processes and Feelings." In *Manuscript Essays and Notes*, edited by Frederick H. Burkhardt, Fredson Bowers, and Ignas K. Skrupskelis. Cambridge, MA: Harvard University Press, 1988.

James, William. "*Du sommeil et des états analogues* by Ambrose A. Liébeault (1868)." In *The Works of William James: Essays, Comments, and Reviews*, edited by Frederick H. Burkhardt, Fredson Bowers, and Ignas K. Skrupskelis. Cambridge, MA: Harvard University Press, 1987.

James, William. "The Energies of Men." In *The Works of William James: Essays in Religion and Morality*, edited by Frederick H. Burkhardt, Fredson Bowers, and Ignas K. Skrupskelis. Cambridge, MA: Harvard University Press, 1982.

James, William. "*Entartung und Genie*, by Cesare Lombroso (1895)." In *The Works of William James: Essays, Comments, and Reviews*, edited by Frederick H. Burkhardt, Fredson Bowers, and Ignas K. Skrupskelis. Cambridge, MA: Harvard University Press, 1987.

James, William. "Fifth Annual Report of the State Board of Health of Massachusetts (1874)." In *The Works of William James: Essays, Comments, and Reviews*, edited by Frederick H. Burkhardt, Fredson Bowers, and Ignas K. Skrupskelis. Cambridge, MA: Harvard University Press, 1987.

James, William. "*Genie und Entartung*, by William Hirsch (1895)." In *The Works of William James: Essays, Comments, and Reviews*, edited by Frederick H. Burkhardt, Fredson Bowers, and Ignas K. Skrupskelis. Cambridge, MA: Harvard University Press, 1987.

James, William. "The Gospel of Relaxation." In *The Works of William James: Talks to Teachers on Psychology and to Students on Some of Life's Ideals*, edited by Frederick H. Burkhardt, Fredson Bowers, and Ignas K. Skrupskelis. Cambridge, MA: Harvard University Press, 1983.

James, William. "G. Papini and the Pragmatist Movement in Italy." *Journal of Philosophy, Psychology and Scientific Methods* 3, no. 13 (1906): 337–41.

James, William. "Great Men and Their Environment." In *The Works of William James: The Will to Believe and Other Essays in Popular Philosophy*, edited by Frederick H. Burkhardt, Fredson Bowers, and Ignas K. Skrupskelis. Cambridge, MA: Harvard University Press, 1979.

James, William. "*Grundzüge der physiologischen Psychologie* by Wilhelm Wundt (1875)." In *The Works of William James: Essays, Comments, and Reviews*, edited by Frederick H. Burkhardt, Fredson Bowers, and Ignas K. Skrupskelis. Cambridge, MA: Harvard University Press, 1987.

James, William. "The Hidden Self." In *The Works of William James: Essays in Psychology*, edited by Frederick H. Burkhardt, Fredson Bowers, and Ignas K. Skrupskelis. Cambridge, MA: Harvard University Press, 1983.

James, William. "Introduction to *The Literary Remains of the Late Henry James*." In *The Works of William James: Essays in Religion and Morality*, edited by Frederick H. Burkhardt, Fredson Bowers, and Ignas K. Skrupskelis. Cambridge, MA: Harvard University Press, 1982.

James, William. "Is Life Worth Living?" In *The Works of William James: The Will to Believe and Other Essays in Popular Philosophy*, edited by Frederick H. Burkhardt, Fredson Bowers, and Ignas K. Skrupskelis. Cambridge, MA: Harvard University Press, 1979.

James, William. "'The Jukes' by Richard. L. Dugdale (1878)." In *The Works of William James: Essays, Comments, and Reviews*, edited by Frederick H. Burkhardt, Fredson Bowers, and Ignas K. Skrupskelis. Cambridge, MA: Harvard University Press, 1987.

James, William. "Lectures on Abnormal Psychology (1895, 1896)." In *The Works of William James: Manuscript Lectures*, edited by Frederick H. Burkhardt, Fredson Bowers, and Ignas K. Skrupskelis. Cambridge, MA: Harvard University Press, 1988.

James, William. "*Lectures on the Elements of Comparative Anatomy, by Thomas Huxley* (1865)." In *The Works of William James: Essays, Comments, and Reviews*, edited by Frederick H. Burkhardt, Fredson Bowers, and Ignas K. Skrupskelis. Cambridge, MA: Harvard University Press, 1987.

James, William. "Lowell Lectures on 'The Brain and the Mind' (1878)." In *The Works of William James: Manuscript Lectures*, edited by Frederick H. Burkhardt, Fredson Bowers, and Ignas K. Skrupskelis. Cambridge, MA: Harvard University Press, 1988.

James, William. "The Medical Advertisement Abomination (1894)." In *The Works of William James: Essays, Comments, and Reviews*, edited by Frederick H. Burkhardt, Fredson Bowers, and Ignas K. Skrupskelis. Cambridge, MA: Harvard University Press, 1987.

James, William. "Medical Registration Act (1894)." In *The Works of William James: Essays, Comments, and Reviews*, edited by Frederick H. Burkhardt, Fredson Bowers, and Ignas K. Skrupskelis. Cambridge, MA: Harvard University Press, 1987.

James, William. "Miscellanea 1: Mostly Concerning Empiricism." In *The Works of William James: Manuscript Essays and Notes*, edited by Frederick H. Burkhardt, Fredson Bowers, and Ignas K. Skrupskelis. Cambridge, MA: Harvard University Press, 1988.

James, William. "The Moral Equivalent of War." In *The Works of William James: Essays in Religion and Morality*, edited by Frederick H. Burkhardt, Fredson Bowers, and Ignas K. Skrupskelis. Cambridge, MA: Harvard University Press, 1982.

James, William. "The Moral Philosopher and the Moral Life." In *The Works of William James: The Will to Believe and Other Essays in Popular Philosophy*, edited by Frederick H. Burkhardt, Fredson Bowers, and Ignas K. Skrupskelis. Cambridge, MA: Harvard University Press, 1979.

James, William. "Notes for a Lecture on 'The Physiological Effects of Alcohol' (1886)." In *The Works of William James: Manuscript Lectures*, edited by Frederick H. Burkhardt, Fredson Bowers, and Ignas K. Skrupskelis. Cambridge, MA: Harvard University Press, 1988.

James, William. "On a Certain Blindness in Human Beings." In *The Works of William James: Talks to Teachers on Psychology and to Students on Some of Life's Ideals*, edited by Frederick H. Burkhardt, Fredson Bowers, and Ignas K. Skrupskelis. Cambridge, MA: Harvard University Press, 1983.

James, William. "The Philosophy of Mental Healing by Leander E. Whipple." In *The Works of William James: Essays, Comments, and Reviews*, edited by Frederick H. Burkhardt, Fredson Bowers, and Ignas K. Skrupskelis. Cambridge, MA: Harvard University Press, 1987.

James, William. "Physiology for Practical Use, ed. by James Hinton (1874)." In *The Works of William James: Essays, Comments, and Reviews*, edited by Frederick H. Burkhardt, Fredson Bowers, and Ignas K. Skrupskelis. Cambridge, MA: Harvard University Press, 1987.

James, William. *A Pluralistic Universe*. Edited by Frederick H. Burkhardt, Fredson Bowers, and Ignas K. Skrupskelis. The Works of William James. Cambridge, MA: Harvard University Press, 1977.

James, William. "*Power through Repose*, by Annie Payson Call (1891)." In *The Works of William James: Essays, Comments, and Reviews*, edited by Frederick H. Burkhardt, Fredson Bowers, and Ignas K. Skrupskelis. Cambridge, MA: Harvard University Press, 1987.

James, William. "The Powers of Men." In *The Works of William James: Essays in Religion and Morality*, edited by Frederick H. Burkhardt, Fredson Bowers, and Ignas K. Skrupskelis. Cambridge, MA: Harvard University Press, 1982.

James, William. *Pragmatism*. Edited by Frederick H. Burkhardt, Fredson Bowers, and Ignas K. Skrupskelis. The Works of William James. Cambridge, MA: Harvard University Press, 1975.

James, William. *The Principles of Psychology*. 3 vols. Edited by Frederick H. Burkhardt, Fredson Bowers, and Ingnas K. Skrupskelis. The Works of William James. Cambridge, MA: Harvard University Press, 1981.

James, William. "*Psychologie der Suggestion*, by Hans Schmidkunz (1892)." In *The Works of William James: Essays, Comments, and Reviews*, edited by Frederick H. Burkhardt, Fredson Bowers, and Ignas K. Skrupskelis. Cambridge, MA: Harvard University Press, 1987.

James, William. "The Psychology of Belief." *Mind* 14, no. 55 (1889): 321–52.

James, William. "*Rapport sur les progrès et la marche de la physiologie générale en France*, by Claude Bernard (1868)." In *The Works of William James: Essays, Comments, and Reviews*, edited by Frederick H. Burkhardt, Fredson Bowers, and Ignas K. Skrupskelis. Cambridge, MA: Harvard University Press, 1987.

James, William. "Rationality, Activity and Faith." *Princeton Review* 2 (1882): 58–86.

James, William. "Recent Works on Mental Hygiene (1874)." In *The Works of William James: Essays, Comments, and Reviews*, edited by Frederick H. Burkhardt, Fredson Bowers, and Ignas K. Skrupskelis. Cambridge, MA: Harvard University Press, 1987.

James, William. "*Responsibility in Mental Disease*, by Henry Maudsley (1874)." In *The Works of William James: Essays, Comments, and Reviews*, edited by Frederick H. Burkhardt, Fredson Bowers, and Ignas K. Skrupskelis. Cambridge, MA: Harvard University Press, 1987.

James, William. "Review of *Human Personality and Its Survival of Bodily Death*, by Frederic W. H. Myers (1903)." In *William James on Psychical Research*, edited by Gardner Murphy and Robert O. Ballou. London: Chatto and Windus, 1961.

James, William. "Scientific View of Temperance (1881)." In *The Works of William James: Essays, Comments, and Reviews*, edited by Frederick H. Burkhardt, Fredson Bowers, and Ignas K. Skrupskelis. Cambridge, MA: Harvard University Press, 1987.

James, William. "The Sentiment of Rationality." In *The Works of William James: The Will to Believe and Other Essays in Popular Philosophy*, edited by Frederick H. Burkhardt, Fredson Bowers, and Ignas K. Skrupskelis. Cambridge, MA: Harvard University Press, 1979.

James, William. "Some Remarks on Spencer's Definition of Mind." In *The Works of William James: Essays in Philosophy*, edited by Frederick H. Burkhardt, Fredson Bowers, and Ignas K. Skrupskelis. Cambridge, MA: Harvard University Press, 1978.

James, William. *Talks to Teachers on Psychology*. New York: Henry Holt and Company, 1946.

James, William. "Talks to Teachers on Psychology." In *The Works of William James: Talks to Teachers on Psychology and to Students on Some of Life's Ideals*, edited by Frederick H. Burkhardt, Fredson Bowers, and Ignas K. Skrupskelis. Cambridge, MA: Harvard University Press, 1983.

James, William. "Two Reviews of *Principles of Mental Physiology*, by William B. Carpenter (1874)." In *The Works of William James: Essays, Comments, and Reviews*, edited by Frederick H. Burkhardt, Fredson Bowers, and Ignas K. Skrupskelis. Cambridge, MA: Harvard University Press, 1987.

James, William. "'Ueber den psychischen Mechanismus hystericher Phänomene' by Josef Breuer and Sigmund Freud (1894)." In *The Works of William James: Essays, Comments, and Reviews*, edited by Frederick H. Burkhardt, Fredson Bowers, and Ignas K. Skrupskelis. Cambridge, MA: Harvard University Press, 1987.

James, William. "*The Unseen Universe* by Peter Guthrie Tait and Balfour Stewart (1875)." In *The Works of William James: Essays, Comments, and Reviews*, edited by Frederick H. Burkhardt, Fredson Bowers, and Ignas K. Skrupskelis. Cambridge, MA: Harvard University Press, 1987.

James, William. *The Varieties of Religious Experience*. Edited by Frederick H. Burkhardt, Fredson Bowers, and Ignas K. Skrupskelis. The Works of William James. Cambridge, MA: Harvard University Press, 1985.

James, William. *The Varieties of Religious Experience: Centenary Edition*. Edited by Eugene Taylor and Jeremy Carrette. London: Routledge 2002.

James, William. "What Is an Emotion?" In *The Works of William James: Essays in Psychology*, edited by Frederick H. Burkhardt, Fredson Bowers, and Ignas K. Skrupskelis. Cambridge, MA: Harvard University Press, 1983.

James, William. "What Makes a Life Significant." In *The Works of William James: Talks to Teachers on Psychology and to Students on Some of Life's Ideals*, edited by Frederick H. Burkhardt, Fredson Bowers, and Ignas K. Skrupskelis. Cambridge, MA: Harvard University Press, 1983.

James, William. "What Psychical Research Has Accomplished (1892)." In *The Works of William James: Essays in Psychical Research*, edited by Frederick H. Burkhardt, Fredson Bowers, and Ignas K. Skrupskelis. Cambridge, MA: Harvard University Press, 1986.

James, William. "What Psychical Research Has Accomplished (1896)." In *The Works of William James: The Will to Believe and Other Essays in Popular Philosophy*, edited by

Frederick H. Burkhardt, Fredson Bowers, and Ignas K. Skrupskelis. Cambridge, MA: Harvard University Press, 1979.

James, William. *The Will to Believe and Other Essays in Popular Philosophy*. Edited by Frederick H. Burkhardt, Fredson Bowers, and Ignas K. Skrupskelis. The Works of William James. Cambridge, MA: Harvard University Press, 1979.

Janet, Pierre. "Les Actes inconscients et la mémoire pendant le somnambulisme." *Revue Philosophique* 25, no. 1 (1888): 238–79.

Janet, Pierre. "Les Actes inconscients et le dédoublement de la personnalité pendant le somnambulisme provoqué." *Revue Philosophique* 22, no. 2 (1886): 577–92.

Janet, Pierre. "L'Anesthésie systématisée et la dissociation des phénomènes psychologiques." *Revue Philosophique* 23, no. 1 (1887): 449–72.

Janet, Pierre. *L'Automatisme psychologique: Essai de psychologie expérimentale sur les formes inférieures de l'activité humaine*. Paris: Alcan, 1889.

Janet, Pierre. "Deuxième note sur le sommeil provoqué à distance et la suggestion mentale pendant l'état somnambulique." *Revue Philosophique* 22, no. 1 (1887): 212–23.

Janet, Pierre. "Note sur quelques phénomènes de somnambulisme." *Revue Philosophique* 21, no. 1 (1886): 190–98.

Janet, Pierre. *Les obsessions et la psychasthénie*. Paris : Félix Alcan, 1903.

Janet, Pierre. "Les Phases intermédiaires de l'hypnotisme." *Revue Scientifique*, 3rd ser., 2 (May 8 1886): 577–87.

Jelliffe. "*Genius and Degeneration* by Dr. William Hirsch." *Journal of Nervous and Mental Diseases* 24, no. 6 (1897): 381–83.

Keyssar, Alexander. *The Right to Vote: The Contested History of Democracy in the United States*. New York: Basic Books, 2009.

Kingsbury, Geo C. "Alcohol: Its Use and Misuse." In *The Humanitarian*, n.d.

Klein, Alexander. "James on Consciousness." In *The Oxford Handbook of William James*, edited by Alexander Klein. Accessed July 7, 2022, https://www.oxfordhandbooks.com /view/10.1093/oxfordhb/9780199395699.001.0001/oxfordhb-9780199395699 -e-4.

Klein, Alexander. "William James's Objection to Epiphenomenalism." *Philosophy of Science* 86, no. 5 (2019): 1179–90.

Koppe, Robert. "Das Alkoholsiechthem und Die Kurzlebigkeit des modernen Menschengeschlechts." Moskau: E. Liessner & J. Romahn, 1894.

Krejci, Mark J. "Self and Healthy-Mindedness: James's *Varieties* and Religion in Psychotherapy." *Streams of William James* 5, no. 1 (2003): 16–20.

Kudlick, Catherine. "Disability History: Why We Need Another 'Other.'" *American Historical Review* 108, no. 3 (2003): 763–93.

Langmore, J. C., and William Roberts. "Reports of Societies." *British Medical Journal*, no. 1 (1865): 494–96.

Lanzoni, Susan. "Sympathy in *Mind*." *Journal of the History of Ideas* 70, no. 2 (2009): 265–87.

Lawlor, Michael S. "William James's Psychological Pragmatism: Habit, Belief and Purposive Human Behaviour." *Cambridge Journal of Economics*, no. 30 (2006): 321–45.

Lecky, William Edward Hartpole. *A History of European Morals from Augustus to Charlemagne*. New York: D. Appleton & Company, 1869.

Lekan, Todd. "Pragmatist Moral Philosophy and Moral Life: Embracing the Tensions." In *The Jamesian Mind*, edited by Sarin Marchetti. London and New York: Routledge, 2022.

Leuret, François. *Du traitement moral de la folie*. Paris: J.-B. Baillière 1840.

Livingston, James. *Pragmatism and the Political Economy of Cultural Revolution, 1850—1940*. Cultural Studies of the United States. Chapel Hill: University of North Carolina Press, 1994.

Lloyd, Brian. *Left Out: Pragmatism, Exceptionalism, and the Poverty of American Marxism, 1890–1920*. Baltimore: Johns Hopkins University Press, 1997.

Lotze, Rudolph Hermann. *Medicinische Psychologie oder Physiologie der Seele*. Leipzig: Wiedmann'sche Buchhandlung, 1852.

Luys, J. *Recherches sur le Système Nerveux Cérébro-Spinal sa Structure, ses Fonctions, ses Maladies*. Paris: J.-B. Bailére et Fils, 1865.

MacDonald, Michael, and Terrence Murphy. *Sleepless Souls: Suicide in Early Modern England*. Oxford Studies in Social History. Paperback ed. Oxford: Clarendon Press, 1993.

Massachusetts Department of Public Health. *State Board of Health of Massachusetts: A Brief History of Its Organization and Its Work, 1869–1912*. Boston: Wright & Potter, 1912.

Matteson, John T. "'Their Facts Are Patent and Startling': WJ and Mental Healing." *Streams of William James* 4, no. 1 (2002): 2–8; 4, no. 2 (2002): 3–7.

Maudsley, Henry. *Body and Mind: An Inquiry into their Connection and Mutual Influence, Specially in Reference to Mental Disorders*. New York: D. Appleton and Company, 1871.

Maudsley, Henry. *Natural Causes and Supernatural Seemings*. 3rd ed. London: Kegan Paul, Trench, Trübner & Co., 1897.

Maudsley, Henry. *Responsibility in Mental Disease*. London: Henry S. King & Co., 1874.

McGranahan, Lucas. *Darwinism and Pragmatism: William James on Evolution and Self-Transformation*. London: Routledge, 2019.

McLellan, J. A. *Applied Psychology: An Introduction to the Principles and Practice of Education*. Toronto: Copp, Clark Company, Ltd., 1890.

Menand, Louis. *The Metaphysical Club*. London: Harper Collins Publishers, 2001.

Merriman, Daniel. "A Sober View of Abstinence." In *Bibliotheca Sacra*. Andover: Warren F. Draper, October 1881.

Mill, John Stuart. *Auguste Comte and Positivism*. London: N. Trübner & Co., 1865.

Mill, J. S. "Utilitarianism." In *Utilitarianism and Other Essays*, edited by Alan Ryan. London: Penguin Books, 1987.

Miller, Joshua I. *Democratic Temperament: The Legacy of William James*. Lawrence: University Press of Kansas, 1997.

Minois, Georges. *History of Suicide: Voluntary Death in Western Culture*. Translated by Lydia G. Cochrane. Baltimore: John Hopkins Press, 1999.

Mitchell, S. Weir, ed. *Five Essays by John Kearsley Mitchell M.D.* Philadelphia: J. B. Lippincott & Co., 1859.

Mitchell, S. Weir. *Wear and Tear, or Hints for the Overworked*. 8th ed. Philadelphia: J. B. Lippincott Company, 1897.

Moll, Albert. *Hypnotism*. London: Walter Scott, 1890.

Montaigne, Michel de. *Essais*. Bordeaux: Simon Millanges, Jean Richer, 1580.

Moore, Gregory. "Nietzsche, Degeneration, and the Critique of Christianity." *Journal of Nietzsche Studies* 19 (Spring 2000): 1–18.

Moran, Maureen. "Hopkins and Victorian Responses to Suffering." *Revue LISA/LISA e-journal* 7, no. 3 (2009): 570–81.

Myers, F. W. H. "First Report of the Literary Committee." *Proceedings of the Society for Psychical Research* 1, no. 2 (1882): 116–55.

Myers, F. W. H. *Human Personality and Its Survival of Bodily Death*. London: Longmans, Green & Co., 1903.

Myers, F. W. H. "Note on a Suggested Mode of Psychical Interaction." In *Phantasms of the Living*, vol. 2, edited by Edmund Gurney, Frederic Myers, and Frank Podmore. Gainesville, FL: Scholars' Facsimiles & Reprints, 1970.

Myers, F. W. H. "On a Telepathic Explanation of Certain So-Called Spiritualistic Phenomena." *Proceedings of the Society for Psychical Research* 2 (1884): 217–37.

Myers, F. W. H. "The Subliminal Consciousness, Chapter 1: General Characteristics of Subliminal Messages." *Proceedings of the Society for Psychical Research* 7 (1892): 298–327.

Myers, F. W. H. "The Subliminal Consciousness, Chapter 2: The Mechanism of Suggestion." *Proceedings of the Society for Psychical Research* 7 (1892): 327–355.

Myers, Gerald E. "Introduction." In *The Works of William James: Talks to Teachers on Psychology and to Students on Some of Life's Ideals*, edited by Frederick H. Burkhardt, Fredson Bowers, and Ignas K. Skrupskelis. Cambridge, MA: Harvard University Press, 1983.

Myers, Gerald E. *William James: His Life and Thought*. New Haven, CT: Yale University Press, 1986.

Myers, Gerald E. "William James's Theory of Emotion." *Transactions of the Charles S. Peirce Society* 5, no. 2 (1969): 67–89.

Neary, Francis. "Interpreting Abnormal Psychology in the Late Nineteenth Century: William James's Spiritual Crisis." In *Histories of the Normal and the Abnormal: Social and Cultural Histories of Norms*, edited by Waltraud Ernst. London and New York: Routledge, 2006.

Nietzsche, Friedrich. *On the Genealogy of Morality*. Translated by Carol Diethe. Edited by Keith Ansell-Pearson. Cambridge: Cambridge University Press, 1994.

Nisard, M., ed. *Oeuvres Complètes de Sénèque le Philosophe, avec la Traduction en Français*. Paris: J. J. Dubochet et Compagnie, 1844.

Nisbet, J. F. *The Insanity of Genius and the General Inequality of Human Faculty Physiologically Considered*. 3rd ed. London: Ward & Downey, 1893.

Nordau, Max. *Degeneration*. London: William Heinemann, 1895.

Nordau, Max. *Entartung*. 2 vols. Berlin: Duncker, 1892–93.

"Notes." *Philosophical Review* 19, no. 6 (1910): 694–97.

Pascal, Blaise. *Les Pensées*. Paris: Alphonse Lemerre, 1777.

Pearce, Trevor. *Pragmatism's Evolution: Organism and Environment in American Philosophy*. London and Chicago: University of Chicago Press, 2020.

Pernick, Martin S. "The Calculus of Suffering in Nineteenth-Century Surgery." *Hastings Center Report* 13, no. 2 (1983): 26–36.

Perry, Ralph Barton. *The Thought and Character of William James: As Revealed in Unpublished Correspondence and Notes, Together with His Published Writings*. Vol. 1, *Inheritance and Vocation*. London: Oxford University Press, 1935.

Perry, Ralph Barton. *The Thought and Character of William James: As Revealed in Unpublished Correspondence and Notes, Together with His Published Writings*. Vol. 2, *Philosophy and Psychology*. London: Oxford University Press, 1935.

Petit, J. B. "Recherches statistiques de l'étiologie du suicide." Paris, 1850.

"The Physiological Influence of Alcohol." *Edinburgh Review*, n.d.

Pick, Daniel. *Faces of Degeneration: A European Disorder, c. 1848–c. 1918*. Cambridge: Cambridge University Press, 1989.

Pickering, W. S. F., and Geoffrey Walford. *Durkheim's Suicide: A Century of Research and Debate*. London: Routledge, 2000.

Pierce, John. "Appendix 2: John Pierce's Summaries in the Journal of Education of James's Lectures." In *The Works of William James: Talks to Teachers on Psychology and to Students on Some of Life's Ideals*, edited by Frederick H. Burkhardt, Fredson Bowers, and Ignas K. Skrupskelis. Cambridge, MA: Harvard University Press, 1983.

Pihlström, Sami. *Taking Evil Seriously*. New York: Palgrave Macmillan, 2014.

Piper, Alta L. *The Life and Work of Mrs. Piper*. London: Kegan Paul, Trench, Trübner & Co., Ltd., 1929.

Podmore, Frank. *Mesmerism and Christian Science: A Short History of Mental Healing*. London: Methuen & Co., 1909.

Porter, Roy. "Nervousness, Eighteenth and Nineteenth Century Style: From Luxury to Labour." In *Cultures of Neurasthenia from Beard to the First World War*, edited by Marijke Gijswijt-Hofstra and Roy Porter. New York: Rodopi B.V., 2001.

Proudfoot, Wayne, ed. *William James and a Science of Religions: Reexperiencing The Varieties of Religious Experience*. New York: Columbia University Press, 2004.

Putnam, James Jackson. "William James." *Atlantic Monthly*, December 1910

Pyle, Andrew. *Utilitarianism: Key Nineteenth-Century Journal Sources*. London: Routledge, 1998.

Quen, J. M. "Isaac Ray and Mental Hygiene in America." *Annals of the New York Academy of Sciences*, no. 291 (1977): 83–93.

Ratcliffe, Matthew. "William James on Emotion and Intentionality." *International Journal of Philosophical Studies* 13, no. 2 (2005): 179–202.

Ray, I. *Mental Hygiene*. Boston: Ticknor and Fields, 1863.

"Review of Fifth Annual Report of the State Board of Health of Massachusetts." *North American Review* 119, no. 245 (1874): 447–52.

Ribot, Th. *The Diseases of Personality*. 2nd ed. Chicago: Open Court Publishing Company, 1895.

Richards, Robert J. "James's Uses of Darwinian Theory." *A William James Renaissance: Four Essays by Young Scholars*, Harvard Library Bulletin 30, no. 4 (1982): 387–425.

Richardson, Robert D. *William James: In the Maelstrom of American Modernism*. New York: Houghton Mifflin Company, 2007.

Robeck, Johannes. *De morte voluntaira philosophorum et bonorum vivorum*, 1735.

Roelcke, Volker. "Electrified Nerves, Degenerated Bodies: Medical Discourses on Neurasthenia in Germany, circa 1880–1914." In *Cultures of Neurasthenia from Beard to the First World War*, edited by Marijke Gijswijt-Hofstra and Roy Porter. New York: Rodopi B.V., 2001.

Rosen, Frederick. *Classical Utilitarianism from Hume to Mill*. London: Routledge, 2003.

Rosenberg, Charles E. "Catechisms of Health: The Body in the Prebellum Classroom." *Bulletin of the History of Medicine* 69, no. 2 (1995): 175–97.

Rousseau, Jean-Jacques. *La nouvelle Héloïse*. Amsterdam: Marc-Michel Rey, 1761.

Rumbarger, John J. *Profits, Power, and Prohibition: Alcohol Reform and the Industrializing of America, 1800–1930*. Albany: State University of New York Press, 1989.

Ruetenik, Tadd. "Fruits of Health; Roots of Despair: William James, Medical Materialism and the Evaluation of Religious Experience." *Journal of Religion and Health* 45, no. 3 (2006): 382–95.

Ruse, Michael, and Robert J. Richards, eds. *The Cambridge Handbook of Evolutionary Ethics*. Cambridge: Cambridge University Press, 2017.

Seneca. *Seneca's Letters to Lucilius*. Vol. 1. Translated by E. Phillips Barker. Oxford: Clarendon Press, 1932.

Shamdasani, Sonu. *Jung and the Making of Modern Psychology: The Dream of a Science*. Cambridge: Cambridge University Press, 2003.

Shapin, Steven, and Christopher Lawrence. "Introduction: The Body of Knowledge." In *Science Incarnate: Historical Embodiments of Natural Knowledge*, edited by Christopher Lawrence and Steven Shapin. Chicago: University of Chicago Press, 1998.

Shook, John R., and Hugh P. McDonald, eds. *F. C. S. Schiller on Pragmatism and Humanism: Selected Writings, 1891–1939*. New York: Humanity Books, 2008.

Simon, Linda. *Genuine Reality: A Life of William James*. New York: Harcourt, Brace and Company, 1998.

Skrupskelis, Ignas K. "Introduction." In *The Works of William James: Manuscript Lectures*, edited by Frederick H. Burkhardt, Fredson Bowers, and Ignas K. Skrupskelis. Cambridge, MA: Harvard University Press, 1988.

Skrupskelis, Ignas K., and Elizabeth M. Berkeley, eds. *The Correspondence of William James*. 12 vols. Charlottesville: University Press of Virginia, 1992–2004.

Small, Miriam. *Oliver Wendell Holmes*. New York: Twayne Publishers Inc., 1963.

Smith, Roger. *Free Will and the Human Sciences in Britain, 1870–1910*. London: Pickering & Chatto, 2013.

Snarey, John, and Paul Jerome Croce, eds. *Streams of William James* 5, no. 3 (2003).

Somburg, Oliver, and Holger Steinberg. "Der Begriff des Seelen-Binnenlebens von Moriz Benedikt." In *Schriftenreihe der Deutschen Gesellschaft für Geschichte der Nervenheilkunde*, edited by W. J. Bock and B. Holdorff, 2006.

Sommer, Andreas. "James and Psychical Research in Context." In *The Oxford Handbook of William James*, edited by Alexander Klein. Accessed May 10, 2021, https://www.oxfordhandbooks.com/view/10.1093/oxfordhb/9780199395699.001.0001/oxfordhb-9780199395699-e-37.

Spencer, Herbert. *Education: Intellectual, Moral and Physical*. New York: D. Appleton and Company, 1861.

Spencer, Herbert. *The Principles of Psychology*. London: Longman, Brown, Green, and Longmans, 1855.

Staël, Madame de. *Réflexions sur le suicide*. London: L. Deconchy, 1813.

Starbuck, Edwin Diller. *The Psychology of Religion: An Empirical Study of the Growth of Religious Consciousness*. London: Walter Scott, Ltd., 1899.

Stephen, Sir Leslie. *The English Utilitarians*. Vol. 1, *Jeremy Bentham*. New York: Augustus M. Kelley, 1968.

Stephens, Piers H. G. "James, British Empiricism, and the Legacy of Utilitarianism." In *The Jamesian Mind*, edited by Sarin Marchetti. London, New York: Routledge, 2022.

Strachan Donnelley, Naomi. "Power and Mysticism in the Introduction of Anesthesia in Nineteenth Century America." MD thesis, Yale University, 1998.

Sutton, Emma K. "Interpreting 'Mind-Cure': William James and the 'Chief Task . . . of the Science of Human Nature.'" *Journal of the History of the Behavioral Sciences* 48, no. 2 (2012): 115–33.

Sutton, Emma. "Marcus Aurelius, William James and the 'Science of Religions.'" *William James Studies* 4 (2009): 70–89.

Sutton, Emma K. "Re-writing 'the Laws of Health': William James on the Philosophy and Politics of Disease in Nineteenth-Century America." PhD thesis, University College London, 2013.

Sutton, Emma K. "When Misery and Metaphysics Collide: William James on 'the Problem of Evil.'" *Medical History* 55, no. 3 (2011): 389–92.

Sweetser, W. *Mental Hygiene, or An Examination of the Intellect and Passions, Designed to Illustrate Their Influence on Health and the Duration of Life.* New York: J. & H. G. Langley, 1843.

Tarver, Erin C., and Shannon Sullivan, eds. *Feminist Interpretations of William James.* University Park: Pennsylvania State University Press, 2015.

Taves, Ann. "The Fragmentation of Consciousness and *The Varieties of Religious Experience*: William James's Contribution to a Theory of Religion." In *William James and a Science of Religions: Reexperiencing The Varieties of Religious Experience*, edited by Wayne Proudfoot. New York: Columbia University Press, 2004.

Taylor, Charles. *Varieties of Religion Today: William James Revisited.* Cambridge, MA: Harvard University Press, 2002.

Taylor, Eugene. "William James and Depth Psychology." In "The Varieties of Religious Experience: Centenary Essays." Special issue, *Journal of Consciousness Studies* 9, no. 9–10 (2002): 11–36.

Taylor, Eugene. *William James on Consciousness beyond the Margin.* Princeton, NJ: Princeton University Press, 1996.

Taylor, Eugene. *William James on Exceptional Mental States: The 1896 Lowell Lectures.* New York: Charles Scribner's Sons, 1982.

Thomas, Joseph M. "Figures of Habit in William James." *New England Quarterly* 66, no. 1 (1993): 3–26.

Thomson, Robert. "Introduction." In William B. Carpenter. *Principles of Mental Physiology, with their Applications to the Training and Discipline of the Mind, and the Study of its Morbid Conditions.* 4th (1876) ed. London: Routledge/Thoemmes Press, 1993.

Townsend, Kim. *Manhood at Harvard: William James and Others.* New York: W. W. Norton & Company, 1996.

Troyer, John, ed. *The Classical Utilitarians: Bentham and Mill.* Indianapolis: Hackett Publishing Company, Inc., 2003.

Tuckey, C. Lloyd. *Psycho-therapeutics, or Treatment by Hypnotism and Suggestion.* 2nd ed. London: Baillière, Tindall, and Cox, 1890.

Tuke, Daniel Hack. *Illustrations of the Influence of the Mind upon the Body in Health and Disease, Designed to Elucidate the Action of the Imagination.* 2nd ed. Philadelphia: Henry C. Lea's Son & Co., 1884.

Tursi, Renee. "William James's Narrative of Habit." *Style* 33, no. 1 (1999): 67–87.

Vande Kemp, Hendrika. "The Gifford Lectures on Natural Theology: Historical Background to James's Varieties." *Streams of William James* 4, no. 3 (2002): 2–8.

Vijselaar, Joost. "Neurasthenia in the Netherlands." In *Cultures of Neurasthenia from Beard to the First World War*, edited by Marijke Gijswijt-Hofstra and Roy Porter. New York: Rodopi B.V., 2001.

Waller, John C. "Ideas of Heredity, Reproduction and Eugenics in Britain, 1800–1875." *Studies in the History and Philosophy of Biology and Bio-medical Sciences* 32, no. 3 (2001): 457–89.

Warren, Joseph W. "Alcohol Again: A Consideration of Recent Misstatements of its Physiological Action." Reprinted from *Boston Medical and Surgical Journal*, July 7, 14, 1887. Boston: Cupples and Hurd 1887.

Warren, Joseph W. "The Effect of Pure Alcohol on the Reaction Time, with a Description of a New Chronoscope." *Journal of Physiology* 8, no. 6 (1887): 341–48.

Westbrook, Robert B. *Democratic Hope: Pragmatism and the Politics of Truth.* Ithaca, NY: Cornell University Press, 2005.

Whipple, Leander Edmund. *The Philosophy of Mental Healing: A Practical Exposition of Natural Restorative Power.* New York: Metaphysical Publishing Company, 1893.

Whittingham Beers, Clifford. *A Mind That Found Itself: An Autobiography.* London, Bombay, Calcutta, and Madras: Longmans, Green and Co., 1917.

Whorton, James C. *Crusaders for Fitness: The History of American Health Reformers.* Princeton, NJ, and Guildford: Princeton University Press, 1982.

Whorton, James C. *Inner Hygiene: Constipation and the Pursuit of Health in Modern Society.* Oxford: Oxford University Press, 2000.

Wiener, Philip P. "Chauncey Wright's Defence of Darwin and the Neutrality of Science." *Journal of the History of Ideas* 6, no. 1 (January 1945): 19–45.

"Wilkinson, James John Garth." In *Dictionary of National Biography, 1885–1900*, 61:271–27. London: Smith, Elder, & Co., 1885–1900.

Wilkinson, James John Garth. *The Greater Origins and Issues of Life and Death.* London: James Speirs, 1885.

Wilkinson, W. M. "The Cases of the Welsh Fasting Girl & Her Father: On the Possibility of Long-Continued Abstinence From Food." London: J. Burns, 1870.

"William James Dies; Great Psychologist." *New York Times*, August 27, 1910.

Winslow, Forbes. *The Anatomy of Suicide.* London: Henry Renshaw, 1840.

Winslow, Forbes. *On Obscure Diseases of the Brain, Disorders of the Mind: Their Incipient Symptoms, Pathology, Diagnosis, Treatment, and Prophylaxis.* Philadelphia: Blanchard and Lea, 1860.

Woody, William Douglas. "Varieties of Religious Conversion: William James in Historical and Contemporary Contexts." *Streams of William James* 5, no. 1 (2003): 7–11.

Wright, Chauncey. "McCosh on Tyndall." In *The Evolutionary Philosophy of Chauncey Wright*, vol. 3, *Influence and Legacy*, edited by Frank X. Ryan. Bristol: Thoemmes Press, 2000.

INDEX

The letter *f* following a page number denotes a figure.

213n48, 214n55, 215n83. *See also* degenerate

Degeneration (Nordau), 138–42, 145, 212n34

demoniacal possession, 142, 204n83

depression, 5, 33. *See also* melancholy

De Quincey, Thomas, 28, 187n65

despair, 25, 115, 127

determinism, 9, 17, 37–39, 42, 182n22

diagnosis, medical: and excavation of memories in trance state, 94, 202n59; of James by others retrospectively, 5; James on, 4, 104, 159, 171; James's self-, 5–7, 45, 107, 127, 139; and mind cure, 91; social consequences of, 4, 30–31, 37–38, 138–46, 171; and spiritualist healers, 85–87

Dietetics of the Soul, The (Feuchtersleben), 59

"Dilemma of Determinism, The" (James), 17, 38, 42

disability studies, 211n6

disease: and criminality, 30–31, 38, 215n83; cultural etiology of, 139, 140; and education, 63–69, 207n36; and epistemology, 153–57; as evil (*see* evil: illness and disease as); excessive drinking as, 55; and habit, 56–57, 58–62, 71; hereditary etiology of, 22, 34, 48–49, 60, 138, 140; James's thought on significance of, 5, 8, 42, 106, 112; mental (*see* insanity); mind-curers on, 89, 91; mortal, 108–14; and mysticism, 154–57; phobia of, 143; physical etiology of, 7, 15, 30, 33–34, 73, 88; prevention of (*see* hygiene; mental hygiene; mental hygiene movement); as punishment from God, 20, 48; relationship with health, 143–44, 172; societal impact of, 4, 47–51; suicidal inclinations as, 30–31; and "unseen world," 150–51, 156–57. *See also* disease, psychological etiology of; illness; *and specific diseases*

disease, psychological etiology of, 7, 35–36, 73, 159–60, 166; and Benedikt's "Binnenleben," 73–74, 119; and heterogeneous personality, 116, 119; in hypochondria and hysteria, 83, 93–94, 95;

and imagination, 96–97, 203n66; and medical advertisements, 97; mind-curers on, 89, 91; and pathological autosuggestion, 95–96; and subconscious self, 10, 84–85, 93–99

disgust, 16, 41, 111; loathing, 115

"divided self," 7, 10, 116–20, 128, 164, 207n46

dogs, 150–51, 157

Donne, John, 28, 186n61

Dorr, George Bucknam, 89, 92, 98, 201n38

dreams, 33, 142

Dresser, Annetta Seabury, 89, 90–92, 201n38

Dresser, Horatio, 90, 201n50

Dreyfus, Alfred, 158

Dumas, Jean, 28

dynamogeny, 122–23, 131

Education (Spencer), 65

Eliot, Charles William, 144, 163

Emerson, Ralph Waldo, 204n96

emotions: and alcohol, 54, 55; buried, 7, 73–74; diseased, 30; and health, 9, 70, 73–74, 75, 77, 173, 197n116; hygienic importance of controlling, 61, 67, 70, 77, 131, 171; as inhibitory notions, 68, 73–74; and insanity, 59, 61, 64, 67, 75, 77; James's theory of, 69–70, 77, 171; and Janet's study of psychasthenia, 127; and muscles, 71–73; and nineteenth-century use of word "moral," 30, 58, 173; and "personal view of life," 81; as philosophical motivations, 8, 35, 42–43, 105; relevance of to medicine, 173; religious, 105, 113, 123, 153; as stimulants, 127–28. *See also specific emotions*

empiricism, 34–37, 189n100

endurance, 75, 122–23, 134; evil and, 40, 42–43, 44, 54, 107–8, 135–36, 169, 172; and war, 8, 129–30

"Energies of Men, The" (James), 10, 126, 128, 130–31, 134, 170, 172

energy: of animal magnetism, 82; "divided self" and wastage of personal, 116–17, 120; emotions as stimulants of, 127–28; ethical and religious faiths as stimulants of, 107, 108, 113, 123; God as

trance state, 82, 85, 87, 98, 203n61,
203n64; and hysteria, 93–94, 202n59,
203n63. *See also* hypnosis
Transcript (newspaper), 51, 158
trauma, 203n63
truth: and disease, 91, 153–57, 169; and
faith, 41–42; and heroism, 151; and observer blindness, 100–102, 215n83; and
philosophy of humanism, 132; and politics, 160; religious, 1, 64, 79–84, 87–
89, 99–106, 121, 169, 170; scientific, 79–
84, 87–89, 99–106, 170
Tuckey, Charles Lloyd, 95
Tuke, Daniel Hack, 64, 96

"unseen world." *See* religion: and "unseen
world" of moral meaning
utilitarianism, 9, 23–27, 185n35, 185n37,
185n40; and suicide, 25–27, 28, 29–
30, 44

Varieties of Religious Experience, The
(James), 1, 173, 206n16, 208n52,
216n102; biographical context of, 10,
108–14, 115, 125–26, 188n85, 205n11; and
death, 108–9, 111–12, 114, 115, 116; and
degeneration, 147–48, 153–55, 213n48;
and disease, 104, 108–14, 134, 153–57;
and "divided self," 10, 116–20, 128; and
energy, 10, 108, 115, 122–23, 125, 128,
129–30, 134, 208n64; and environment,
adaptation and "correspondence" to,
124, 149, 169; and evil, 17, 108, 113, 114,
115–16, 124, 129–30, 134, 155–57, 169;
and fighting temper, 128, 129–30; and
God, 108, 114, 120, 121–22, 123–25, 134;
and health, 10, 104, 115, 122–25, 134,
173; and "healthy-minded" and "sick-
souled" people, 115–16, 124, 155–56,
169; heterogeneous personality, 116–17,
119; and inhibition, 69, 117–18, 119–
20, 125, 128, 129–30; James's philosophy of pragmatism, 102–4, 154, 173;
and melancholy, 115, 123, 124, 156, 157;
and mystical insight, 153–57, 169; and
regeneration, 104, 120, 124, 128; and
religious conversion, 10, 117–18, 119–
22, 207–8nn48–49; and saintliness,
148–49, 152; and spiritual heroism,

10, 147, 149, 152; and Stoicism, 112–14,
134; and subconscious, 118–20, 121–23,
208n57; and will, 113, 114, 118–20, 130,
134, 208n49
vis medicatrix Naturæ, 123, 208n65
vitality, 115, 122, 127, 129, 147–48, 172
volition. *See* will

war, 6, 148; and energy and endurance, 8,
128–30; against evil, 21–22, 43, 44, 128,
131, 152, 156–57, 214n59; internal, 116,
164; against nature, 128–29. *See also*
American Civil War; "Moral Equivalent of War, The" (James)
Ward, Thomas Wren, 14
Wear and Tear (Mitchell), 64, 129
West Malvern (England), 109
What Katy Did (Coolidge), 149
"What Makes a Life Significant"
(James), 40, 135–36
"What Psychical Research Has Accomplished" (James), 105
Whipple, Leander, 98
Wilkinson, James John Garth, 3, 87–89
will, 58, 91, 173, 190n110, 195n69; and
emotional control, 59, 61, 70, 131, 171;
and energy, 130–31, 134; and illness,
96, 130, 131, 134; and inhibition, 68–
69; and insanity, 30, 61–62; as muscle, 67–68; pathologies of, 30, 118–20,
207n43; as psychologically redundant
concept, 118, 207n44; and religious
self-surrender, 114, 134; training of, 59,
67–68, 130–31, 134. *See also* free will;
self-control; "Will to Believe, The"
(James)
"Will to Believe, The" (James), 1, 40,
146–47, 150
Winslow, Forbes, 30, 31–33, 32f, 188n83
worry, 53, 75–77, 213n37; anxiety, 71, 76–
77, 115
Wright, Chauncey, 34–35, 189n96, 189n98,
189n100, 198n2

yoga, hatha, 130–31, 133, 170

Zola, Émile, 139, 158
Zur Genealogie der Moral (Nietzsche),
147–48, 214n55